Hack Audio

An Introduction to Computer Programming and Digital Signal Processing in MATLAB®

Computers are at the center of almost everything related to audio. Whether for synthesis in music production, recording in the studio, or mixing in live sound, the computer plays an essential part. Audio effects plug-ins and virtual instruments are implemented as software computer code. Music apps are computer programs run on a mobile device. All these tools are created by programming a computer.

Hack Audio: An Introduction to Computer Programming and Digital Signal Processing in MATLAB provides an introduction for musicians and audio engineers interested in computer programming. It is intended for a range of readers including those with years of programming experience and those ready to write their first line of code. In the book, computer programming is used to create audio effects using digital signal processing. By the end of the book, readers implement the following effects: signal gain change, digital summing, tremolo, auto-pan, mid/side processing, stereo widening, distortion, echo, filtering, equalization, multi-band processing, vibrato, chorus, flanger, phaser, pitch shifter, auto-wah, convolution and algorithmic reverb, vocoder, transient designer, compressor, expander, and de-esser.

Throughout the book, several types of test signals are synthesized, including sine wave, square wave, sawtooth wave, triangle wave, impulse train, white noise, and pink noise. Common visualizations for signals and audio effects are created including waveform, characteristic curve, goniometer, impulse response, step response, frequency spectrum, and spectrogram. In total, over 200 examples are provided with completed code demonstrations.

Eric Tarr is an assistant professor of Audio Engineering Technology at Belmont University in Nashville, TN. He teaches classes on digital audio, computer programming, and signal processing. He received a Ph.D., M.S., and B.S. in Electrical and Computer Engineering from the Ohio State University. He received a B.A. in Mathematics and a minor in Music from Capital University in Columbus, OH. His work has spanned across the topics of speech signal processing, musical robotics, sound spatialization, acoustic and electronic system modeling, hearing loss, perception, and cognition. He has published articles in the *Journal of the Acoustical Society of America*, *Journal of Speech, Language, and Hearing Research*, *International Journal of Audiology*, and *Mechanical Engineering Research*. He is a member of the Audio Engineering Society. In 2015, Dr. Tarr received the Gibson Foundation Les Paul Music Innovation Award as the principal investigator of a research grant on "Blockchain Technology in the Music Industry."

AUDIO ENGINEERING SOCIETY PRESENTS...

http://www.aes.org/

Editorial Board

Chair: Francis Rumsey, Logophon Ltd.
Hyun Kook Lee, University of Huddersfield
Natanya Ford, University of West England
Kyle Snyder, Ohio University

Other titles in the Series:

Handbook for Sound Engineers, 5th Edition
Edited by Glen Ballou

Audio Production and Critical Listening, 2nd Edition
Authored by Jason Corey

Recording Orchestra and Other Classical Music Ensembles
Authored by Richard King

Recording Studio Design, 4th Edition
Authored by Philip Newell

Modern Recording Techniques, 9th Edition
Authored by David Miles Huber

Immersive Sound: The Art and Science of Binaural and Multi-Channel Audio
Edited by Agnieszka Roginska and Paul Geluso

Hack Audio: An Introduction to Computer Programming and Digital Signal Processing in MATLAB®
Authored by Eric Tarr

Hack Audio

*An Introduction to Computer Programming and
Digital Signal Processing in MATLAB®*

Eric Tarr

Routledge
Taylor & Francis Group

LONDON AND NEW YORK

First published 2019
by Routledge
711 Third Avenue, New York, NY 10017

and by Routledge
2 Park Square, Milton Park, Abingdon, Oxon, OX14 4RN

Routledge is an imprint of the Taylor & Francis Group, an informa business

MATLAB® and Simulink® are registered trademarks of The MathWorks, Inc. and are used with permission. The MathWorks does not warrant the accuracy of the text or exercises in this book. This book's use or discussion of MATLAB® and Simulink® software or related products does not constitute endorsement or sponsorship by the MathWorks of a particular pedagogical approach or particular use of the MATLAB® and Simulink® software.

For MATLAB and Simulink product information, please contact:
The MathWorks, Inc.
3 Apple Hill Drive
Natick, MA 01760-2098 USA
Tel: 508-647-7000
Fax: 508-647-7001
E-mail: info@mathworks.com
Web: mathworks.com
How to buy: www.mathworks.com/store

Trademark notice: Product or corporate names may be trademarks or registered trademarks, and are used only for identification and explanation without intent to infringe.

Library of Congress Cataloging-in-Publication Data

Names: Tarr, Eric, author.
Title: Hack audio : an introduction to computer programming and digital signal processing in MATLAB / Eric Tarr.
Description: New York, NY : Routledge, 2019. | Series: Audio Engineering Society presents
Identifiers: LCCN 2018006035| ISBN 9781138497542 (hardback : alk. paper) | ISBN 9781138497559 (pbk. : alk. paper) | ISBN 781351018463 (ebook)
Subjects: LCSH: MATLAB. | Computer sound processing.
Classification: LCC ML74.4.M36 T37 2018 | DDC 006.5–dc23
LC record available at https://lccn.loc.gov/2018006035

ISBN: 978-1-138-49754-2 (hbk)
ISBN: 978-1-138-49755-9 (pbk)
ISBN: 978-1-351-01846-3 (ebk)

Typeset in Times
by diacriTech

Dedicated to

my supportive wife, Kate

Contents

List of Tables

List of Figures

List of Examples

Acknowledgements

I would like to acknowledge the contributions of my students in the development of this book. It would not have been written without their enthusiasm and feedback. I would like to give a special thanks to Paul Mayo for helping to revise the book's content.

Chapter 1

Introduction

1.1 Introduction: Computer Programming and Digital Signal Processing

Computers are powerful tools for recording, analyzing, visualizing, and processing digital audio. Programming a computer allows for the robust control of audio signals and the creation of software for specialized tasks. Digital signal processing (DSP) is an extensive field of mathematical methods with many applications to process digital audio signals. Together, computer programming and DSP provide a framework for accomplishing many audio-related tasks.

1.2 The Purpose of This Book

The purpose of this book is to provide an introduction for audio engineers to computer programming and digital signal processing. These topics are presented generally, and also demonstrated using the MATLAB® programming language. Furthermore, many pertinent features of the MATLAB Signal Processing Toolbox™ are covered.

The content of the book is intended to supplement a more general introduction to programming by an undergraduate computer science course and introduction to DSP by an undergraduate electrical engineering course. However, topics are presented for readers with little or no experience with computer programming or DSP. Therefore, this book can be used as a precursor for more general study. Additionally, the material in this book is intended to provide a foundation for advanced study specifically in audio engineering.

1.3 Intended Readers

This book is intended for students who would like to learn about computer programming through the use of examples and applications in audio. A mathematical foundation at the level of college algebra and trigonometry is expected. Consequently, audio signal processing can be an intuitive context to practically apply mathematical theory. It is not necessary to have previous knowledge of calculus because all processing is performed with digital signals. Additionally, it is helpful to have a foundation in the physics of sound. Necessary topics include a foundational understanding of signals, vibration, amplitude, frequency, and sampling.

Additionally, this text is intended for practicing engineers needing a handbook to reference with a wide range of implemented examples and detailed background explanations.

1.4 Topics Covered

The following is a list of topics covered in the book. An introduction to programming using the MATLAB language is presented in Chapter 2. The basics of working with audio signals are discussed in Chapter 3, along with common methods to visualize audio. An overview of the MATLAB programming environment is provided in Chapter 4. The programming constructs of several control structures are presented in Chapter 5. In Chapter 6, the basics of amplitude processing are discussed, including gain scaling and DC offset. The synthesis of common audio test signals is covered in Chapter 7. From there, the topics pivot to applying computer programming to perform DSP algorithms. Chapter 8 presents digital summing, signal fades, and amplitude modulation. Chapters 9 and 10 cover stereo audio and nonlinear distortion effects, respectively. Chapters 11, 12, and 13 introduce systems with delay processing to create echo effects and spectral effects. Circular buffers and fractional delay are covered in Chapter 14. Modulated delay effects are demonstrated in Chapter 15, and these modulation effects are used as the building blocks to create algorithmic reverberation in Chapter 16. Finally, effects based on a signal's amplitude envelope are covered in Chapter 17 and become the foundation of dynamic range effects in Chapter 18.

The topics are organized sequentially based on requisite computer programming knowledge and by increasing complexity for signal processing. However, the presentation of material is intended to allow for topical study for readers with prior computer programming and signal processing experience.

1.5 Additional Content

To supplement the material in the book, there is *Additional Content* available at the *Hack Audio* website (www.hackaudio.com). Readers can access MATLAB scripts, functions, and sound files that accompany various examples. It is highly recommended that students use the *Additional Content* to follow along with the content of the book.

Chapter 2

Basics of Programming in MATLAB®

2.1 Introduction: Computer Programming in MATLAB

Computers can be used for calculations, analysis, visualization, and processing. These are all important tasks when working with audio. In order to program a computer to complete all these tasks with audio information, it is important to understand how a computer performs these tasks with more general types of information. This chapter introduces necessary programming concepts in preparation for working with audio in Chapter 3.

2.2 Programming Languages

A **programming language** is a vocabulary or a set of instructions a computer can interpret. Computers can interpret many different programming languages. Despite their differences, many languages have similar constructs. As an analogy, there are many spoken languages that use the same conceptual constructs (nouns, verbs, punctuation, etc.), yet might incorporate the constructs differently. The rules for what is allowed in a language is called **syntax**.

Different programming languages have various advantages and disadvantages. Some languages allow for advanced control of the computer, but are complex. Other languages are simpler, but have less control. MATLAB is a programming language specialized for scientific computing that makes many mathematical and engineering tasks intuitive while maintaining advanced control of the computer. It is used in a diverse range of fields including statistics, finance, biomedical engineering, computer vision, wireless communication, robotics, and economics. More specifically for audio, MATLAB is a common language for DSP.

Examples of other general-purpose programming languages are C/C++, Java, HTML, Python, Fortran, Visual Basic, Swift, Objective-C, assembly, and machine code. Comparably, MATLAB has many aspects that make it very useful for working with audio. There are other programming languages created specifically for working with audio. Some examples are Csound, FAUST, Max/MSP, ChucK, Pure Data, and SuperCollider.

2.3 *Executed Commands*

A computer executes commands provided by a programmer to perform tasks. There are several analogies that can be used to illustrate the idea of executed commands by a computer: architectural blueprints provide the steps for a construction crew to build a structure; the lines in a movie script are followed by actors and actresses to perform a scene; and a recipe is followed by bakers and chefs when completing a dish or meal.

Just as there is an order to how blueprints, scripts, and recipes are meant to be completed, a computer executes commands in a particular order. Therefore, it is important to carefully organize commands in the proper order for an application.

A command can be executed in MATLAB by typing it in the **Command Window** and pressing [RETURN].

2.3.1 *Error Statements*

If a command is entered that cannot be executed by the computer, an **error statement** is displayed. The contents of the error statement depend on the problem with the executed command. The topic of error statements is further explored throughout this book.

2.4 *Mathematics*

One task computers excel at performing is mathematics. Even basic calculators are computers that execute mathematical commands. Conventionally, computers use numbers and operators for mathematics.

2.4.1 *Operators*

Many common mathematical symbols are designated as **operators** in programming languages. MATLAB uses the following symbols for basic mathematical operations: + (addition), - (subtraction), * (multiplication), / (division), and ^ (exponentiation). Parantheses, (and), are also available in MATLAB to control the order of operations.

MATLAB executes calculations based on an order of operations used in most calculators. A programmer should be intentional about the order executed in their commands. Otherwise, unexpected results from the computer may occur. The order of precedence from highest to lowest is: parentheses (for nested parentheses, the innermost are executed first), exponentiation, multiplication and division (equal precedence), addition and subtraction (equal precedence). If an expression has multiple operations of equal precedence, the expression is executed from left to right.

In Example 2.1, several commands are provided to demonstrate the use of mathematical operators. Each command should be executed individually and the answer observed.

Examples: Execute the following commands in MATLAB.

```
>> 2 + 3
>> 2 - 3
>> 2 * 3
>> 2 / 3
>> 10 + 8 / 2
>> (10 + 8) / 2
>> 2 ^ 3
```

Example 2.1: Arithmetic in MATLAB

There are other mathematical operations for which an operator symbol is not defined. Instead, MATLAB has many predefined mathematical functions, called **built-in functions**. These functions can be used along with mathematical operators. Table 2.1 displays several common mathematical functions in MATLAB. More information can be found in the MATLAB help documentation for *functions*.

In Example 2.2, there are several commands that demonstrate the use of mathematical functions. Each command should be executed individually and the answer observed.

Table 2.1: Mathematical operators in MATLAB

`sqrt(x)`	Square Root	`log2(x)`	Base 2 Logarithm log_2
`sin(x)`	Sine	`max(x)`	Maximum Value
`asin(x)`	Arc Sine	`min(x)`	Minimum Value
`cos(x)`	Cosine	`abs(x)`	Absolute Value
`acos(x)`	Arc Cosine	`sign(x)`	Signum Function
`tan(x)`	Tangent	`round(x)`	Round to Nearest Integer
`atan(x)`	Arc Tangent	`fix(x)`	Round towards 0
`exp(x)`	Natural Number e^x	`ceil(x)`	Round towards $+\infty$
`log(x)`	Natural Logarithm log_e	`floor(x)`	Round towards $-\infty$
`log10(x)`	Base 10 Logarithm log_{10}	`conj(x)`	Complex Conjugate

Examples: Execute the following commands in MATLAB.

```
>> sqrt(9)
>> abs(-9)
>> log10(100)
>> cos(0)
>> asin(1)
>> exp(1)
>> sign(5)
>> sign(-5)
>> round(8.1)
>> ceil(8.1)
```

Example 2.2: Mathematical functions in MATLAB

2.4.2 Variables

Variables are an aspect of computer programming languages that allow for information to be assigned, stored, and recalled. As an example, when MATLAB is used as a basic calculator, the result of a command is stored in the MATLAB Workspace as the current answer, called ans. This result can be iteratively recalled. In Example 2.3, notice that the result from the first executed command is used as the value of the variable ans in the second command. Then, the result of the second command is used as the value of the variable ans in the third command.

Examples: Execute the following commands in MATLAB.

```
>> 5 + 6
>> ans + 3
>> ans + 4
```

Example 2.3: Using the ans variable

During the execution of this command, MATLAB has implicitly assigned the result of the calculation to a **variable**. Variables can be explicitly created using an equals sign, =. This is more accurately called the **assignment operator** in computer programming. When a variable is created, it is assigned a value and given a name. Then, when the variable name is used in subsequent commands, the variable value is inserted in place of the variable name.

The process to create a variable is the following. As part of a command, the name of a variable is written by itself on the left side of the assignment operator. The variable is assigned the value of the result of the right side of assignment operator. It should be noted that the reverse order is not permitted in MATLAB. In Example 2.4, the syntax for assigning variables is demonstrated. An example of using invalid syntax is included to demonstrate an error to avoid.

Examples: Execute the following commands in MATLAB.

```
>> a = 1
>> x = 5 + 6
>> y = x - 3
>> y + x
>> zErr = 3x
>> z = 3*x
```

Example 2.4: Assigning variables

Just as a value can be assigned or written to a variable, it can also be overwritten. Therefore, it is always important to be aware of how the value of a variable can change throughout a computer program.

Examples: Execute the following commands in MATLAB.

```
>> h = 4
>> y = h + 1
>> h = y
>> y = h + 1
```

Example 2.5: Overwriting variables

A semicolon can be used at the end of an assignment statement to suppress the output to the MATLAB Command Window. In other words, the result of the assignment statement is not printed to the Command Window, yet the variable is still stored in the Workspace.

Examples: Execute the following commands in MATLAB.

```
>> r = 12;
>> t = r - 6;
>> s = r / t
```

Example 2.6: Suppressing execution output

Variable Naming Rules and Conventions

The names of variables can be individual letters or multiple letters. Variable names can also contain numbers, but the first character in a variable name cannot be a number. Other symbols (@, #, $, %, &, etc.) commonly found on QWERTY keyboards cannot be used in variable names. Variable names cannot contain spaces.

It is common practice to name variables based on their meaning or purpose whenever possible. Although the naming of variables is arbitrary from the computer's perspective, it can help a programmer understand command statements in a program. Furthermore, many programmers follow a convention called *camel case* or *camel caps*. In this convention, variable names start with lowercase letters. Then, the first letter in concatenated words is capitalized (e.g., `numOfSeconds`, `audioSignal`). In Example 2.7, the naming rules and conventions of variables are demonstrated, including examples of invalid names.

Examples: Execute the following commands in MATLAB.

```
>> regulationTime = 90;
>> extraTime = 4;
>> totalTime = regulationTime + extraTime;
>> 4square = 66;
>> square4 = 66;
>> four square = 66;
```

Example 2.7: Naming variables

2.5 *Data Types*

Computers can process different types of data or information. Just as many spoken and written languages use different types of information, computer languages also make use of various data types. As an example, the number 2 in the English language can be written with a numerical symbol, "2". It can also be written with a sequence of text characters, "two". In certain circumstances it is appropriate to use the numerical symbol and in other circumstances it is appropriate to use the text characters. A computer processes and stores information on different data types depending on when it is appropriate, efficient, and intuitive to a user.

2.5.1 *Numbers*

Programming languages, including MATLAB, can store numerical representations of information or data. An example of this is a variable that stores a number as its value. Another name for a single number is a **scalar**.

There are many different types of numerical representations in MATLAB including 16-bit fixed-point, 24-bit fixed-point, and 32-bit floating-point numbers, which are common for digital audio. As a default, MATLAB uses 64-bit double-precision floating-point numbers. Although there are many different types, in many cases it is not necessary for a MATLAB programmer to determine which to use, as it is automatically set by the computer.

2.5.2 Characters

Aside from storing numbers as values, MATLAB can store text as information. A single unit of text is called a **character**. Characters can be letters, symbols, and numbers. Characters can be assigned to variables, and created by putting text between single quotation marks.

Examples: Execute the following commands in MATLAB.

```
>> Y = 'a'
>> X = 'A'
>> W = '#'
>> errW = #
>> V = '3'
```

Example 2.8: Creating characters

It is important to note that the use of a numerical symbol as a character has a different meaning to a computer than the same symbol as a numerical data type (i.e., '3' is different than 3).

2.5.3 Strings

Multiple characters can be combined together to create a **string**. Letters, symbols (including white space), and numbers can be used in the same string by enclosing the characters in single quotations.

Examples: Execute the following commands in MATLAB.

```
>> Y = 'aA#3'
>> X = 'Hello World!'
```

Example 2.9: Creating strings

String Concatenation

Two or more strings can be combined together to create a new string. This process is called **concatenation**, and can be accomplished by using square brackets, [,], or the built-in function strcat.

Examples: Execute the following commands in MATLAB.

```
>> a = 'Hello'
>> b = 'World'
>> c = [ a b ]
>> cc = strcat(a,b)
>> B = ' World!'
```

```
>> C = [ a B ]
>> d = [ a ; b ]
>> D = [ a ; B ]
>> D = char(a,b)
```

Example 2.10: String concatenation

2.6 Arrays

In MATLAB, **arrays** are used to arrange and group multiple values together. The individual values in an array are called **elements**. If strings are a method to arrange and group characters, arrays are a more general method of grouping information.

Arrays can contain sets of numbers, characters, strings, or even other arrays. Just as it is with strings, the process of combining multiple arrays is called concatenation. As a rule, a single array is limited to a single data type. An array cannot contain different data types.

The elements in an array must be arranged in a rectangular shape with one or more columns and one or more rows. To be a valid array, if one column has a certain number of rows, then all other columns must have the same number of rows. Likewise, if one row has a certain number of columns, then all other rows must have the same number of columns.

2.6.1 Basic Array Creation

A one-dimensional array, or **vector**, is a sequence of information in either one row or one column. A row vector is created by enclosing a sequence of data, separated by spaces, between square brackets. A column vector is created by enclosing a sequence of data, separated by semicolons, between square brackets.

Examples: Execute the following commands in MATLAB.

```
>> arr = [ 5 ]
>> rowVec = [ 1 2 3 4 ]
>> colVec = [ 1 ; 2 ; 3 ; 4 ]
>> charVec = [ 'H' 'e' 'l' 'l' 'o' ]
>> strCol = [ 'One' ; 'Two' ]
>> strErr = [ 'One' ; 'Two' ; 'Three' ]
>> strRow1 = [ 'One' 'Two' 'Three']
>> strRow2 = [ 'One ' 'Two' ' Three']
>> rowConcat = [ rowVec arr ]
>> colErr = [ colVec arr ]
>> colConcat = [ colVec ; arr ]
```

Example 2.11: Creating arrays

A two-dimensional array, or matrix, is a set of information stored in multiple rows and columns. The dimensions of a matrix are rectangular. Therefore, it is not necessary for the number of rows to be the same as the number of columns.

Examples: Execute the following commands in MATLAB.

```
>> mat1 = [ 1 2 3 ; 4 5 6 ]
>> mat2 = [ 1 2 ; 3 4 ; 5 6]
>> mat3 = [ mat1 ; 7 8 9 ]
>> mat4 = [ mat2 ; 7 8 ]
```

Example 2.12: Creating matrices

2.6.2 *Plotting Arrays*

The MATLAB function, `plot`, can be used for plotting arrays. By default, it plots a line connecting the values of an array. With one input variable, the function, `plot(t)`, displays a line connecting the values of each element, $t[n]$, versus the element number, n.

Examples: Execute the following commands in MATLAB.

```
>> t = [ 5 2 4 4 -3 3 ];
>> plot(t);
```

Example 2.13: Plotting arrays

The resulting output is shown in Figure 2.1. In this case, the value of each element is plotted along the vertical axis versus the element number on the horizontal axis.

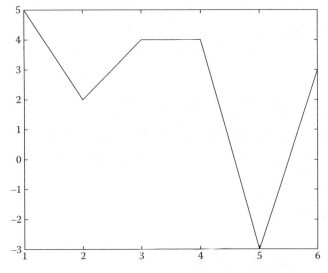

Figure 2.1: Example of the `plot` function

2.7 *Mathematical Functions*

A mathematical function is a relationship between a dependent variable (y) and an independent variable (x). The value of "y" is based on, or dependent on, the value of "x." As an example, a linear relationship between a dependent variable and independent variable is $y = mx + b$, where "m" refers to the slope and "b" refers to the offset or y-intercept.

For every value of "x," a value of "y" can be calculated. Given a value of "x," the value of "y" is written $y(x)$. A sequence of values of "y" at regular intervals, $x = \{1, 2, 3, 4, ...\}$, can be written $y(x) = \{y(1), y(2), y(3), y(4), ...\}$. A MATLAB array is one method to store the values of mathematical functions.

2.7.1 *Plotting Mathematical Functions*

With two input variables, the function plot(x,y) displays a line connecting the values of the second variable, y, versus the first variable, x. In Example 2.14, the provided commands can be executed to assign arrays to variables. Then, the variables are plotted to display several values of a mathematical function.

Examples: Execute the following commands in MATLAB.

```
>> m = 3;
>> b = 1;
>> x = [ 0 1 2 3 ];
>> y = [ (m*0+b) , (m*1+b) , (m*2+b) , (m*3+b) ]
>> plot(x,y);
>> t = [ 0 , pi/2 , pi , 3*pi/2 , 2*pi];
>> a = [ sin(0) , sin(pi/2) , sin(pi) , sin(3*pi/2) , sin(2*pi)];
>> plot(t,a);
```

Example 2.14: Plotting mathematical functions

2.8 *APPENDIX: Additional Plotting Options*

There are many plotting options in MATLAB for customizing the visualization of information. This appendix covers several of the options. The MATLAB help documentation for plot provides complete details of available functionality.

2.8.1 *Line Specification*

There are many possibilities for setting the display options of a line on a plot. These options include the color, style, and width of a line.

Table 2.2: Plot line colors and styles in MATLAB

Specifier	Color	Specifier	Style
'r'	Red	'-'	Solid Line (Default)
'g'	Green	'--'	Dashed Line
'b'	Blue	':'	Dotted Line
'c'	Cyan	'-.'	Dash-Dot Line
'm'	Magenta	'o'	Circle Marker
'y'	Yellow	'*'	Asterisk Marker
'k'	Black	'+'	Plus Sign Marker
'w'	White	'.'	Dot Marker

Line Color and Style

The color and style for each line can be set in the plot function. Table 2.2 displays several options for additional input variables to the plot function.

The **line specifiers** for color and style can be used together in the same set of single quotations as a single input variable to the plot function; this is demonstrated in Example 2.15. The MATLAB documention for *Line Specification* provides additional details about line options.

Line Width

The width of the plotted line can be set using the 'LineWidth' attribute. It is specified as an input variable to the plot function, followed by a positive number. The 'LineWidth' attribute is demonstrated in Example 2.15, and the result is shown in Figures 2.2 and 2.3.

Examples: Execute the following commands in MATLAB.

```
>> x = [ 0 ; 1 ; 2 ; 3 ; 4 ];
>> y = [ 0^2 ; 1^2 ; 2^2 ; 3^2 ; 4^2 ];
>> plot(x,y,'r--','LineWidth',4);
>> z = [ 0^3 ; 1^3 ; 2^3 ; 3^3 ; 4^3 ];
>> plot(x,z,'k:o','LineWidth',2);
```

Example 2.15: Plot line options

Stem Plot

A similar function to plot is the stem function, which produces a stem plot. An example of the stem function is shown in Figure 2.4.

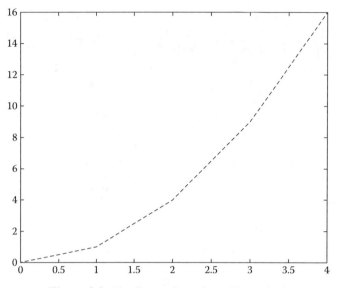

Figure 2.2: Plot line style options: Example 1

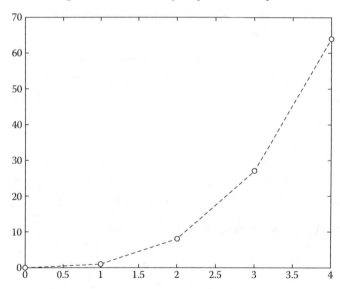

Figure 2.3: Plot line style options: Example 2

Examples: Execute the following commands in MATLAB.

```
>> x = [ 0 ; 1 ; 2 ; 3 ; 4 ];
>> y = [ sqrt(0) ; sqrt(1) ; sqrt(2) ; sqrt(3) ; sqrt(4) ];
>> stem(x,y)
```

Example 2.16: Stem plot function

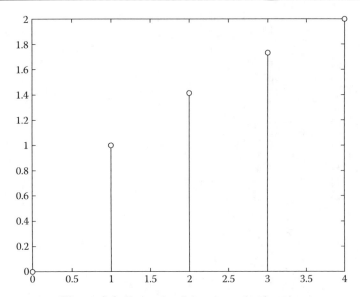

Figure 2.4: Example of the `stem` plot function

2.8.2 Axis Labels

The label of the *x*- and *y*-axes can be set by using the functions `xlabel('string')` and `ylabel('string')`, respectively. The use of these functions is demonstrated in Example 2.17.

2.8.3 Figure Title

The title of a figure can be set by using the function `title('string')`. The use of this function is demonstrated in Example 2.17.

2.8.4 Figure Legend

A legend can be added to a figure by using the function `legend('string1','string2', ...)`. It is necessary for the number of input variables to the legend function to match the number of plotted arrays. The use of this function is demonstrated in Example 2.18.

2.8.5 Axis Scale

The axis scale in a figure can be changed from linear to logarithmic by using the following functions instead of `plot(x,y)`. The function `semilogx(x,y)` makes the *x*-axis logarithmic. The function `semilogy(x,y)` makes the *y*-axis logarithmic. The function

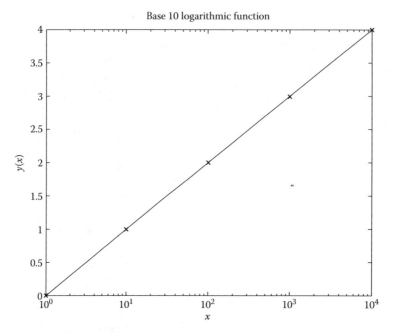

Figure 2.5: Example of the semilog plot function

loglog(x,y) makes the *x*- and *y*-axes logarithmic. In Example 2.17, several plotting functions are demonstrated with the result shown in Figure 2.5.

Examples: Execute the following commands in MATLAB.

```
>> x = [ 1 ; 10 ; 100 ; 1000 ; 10000 ];
>> y = [log10(1);log10(10);log10(100);log10(1000);log10(10000)];
>> semilogx(x,y,'-x');
>> title('Base 10 Logarithmic Function');
>> xlabel('x');
>> ylabel('y(x)');
```

Example 2.17: Logarithmic axis scale

2.8.6 Axis Dimensions

By default, the span of the *x*-axis and *y*-axis in a figure fits the data. However, the span of the *x*-axis and *y*-axis can be user specified by using the axis function. The following syntax can be used to set the limits for the *x*-axis and *y*-axis on the current figure: axis([xmin xmax ymin ymax]). Example 2.18 demonstrates the function to change the axis dimensions with the result shown in Figure 2.6.

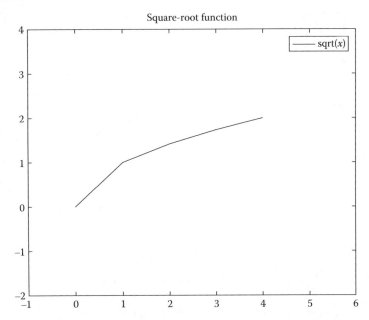

Figure 2.6: Example of the axis and legend functions

Examples: Execute the following commands in MATLAB.

```
>> x = [ 0 ; 1 ; 2 ; 3 ; 4 ];
>> y = [ sqrt(0) ; sqrt(1) ; sqrt(2) ; sqrt(3) ; sqrt(4) ];
>> plot(x,y);
>> title('Square-Root Function');
>> axis([-1 6 -2 4]);
>> legend('sqrt(x)');
```

Example 2.18: Setting axis dimensions

2.8.7 Multiple Arrays

There are several possible methods to plot multiple arrays. Multiple figure windows can be opened and each set with their own plot. Multiple arrays can be plotted on the same axes in a figure. Additionally, separate axes can be plotted within the same figure.

Multiple Windows

The figure function can be used to open and plot in multiple windows. By default, each figure window is automatically numbered. A plot can be placed in a specified window by including a number as an input variable to the figure(num) function. The use of this function is demonstrated in Example 2.19.

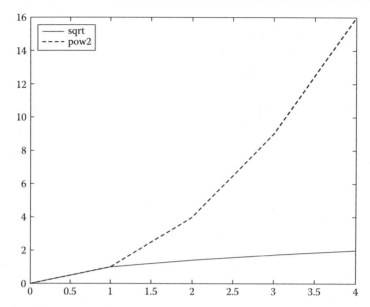

Figure 2.7: Plot of two functions in one figure window

Identical Dimensions

Multiple arrays can be plotted in the same figure by using the `plot` function, if the dimensions of the arrays are identical. This is demonstrated in Example 2.19 and Figure 2.7.

Examples: Execute the following commands in MATLAB.

```
>> x = [ 0 ; 1 ; 2 ; 3 ; 4 ];
>> y = [ sqrt(0) ; sqrt(1) ; sqrt(2) ; sqrt(3) ; sqrt(4) ];
>> z = [ 0^2 ; 1^2 ; 2^2 ; 3^2 ; 4^2 ];
>> plot(x,y);
>> figure;
>> plot(x,z);
>> figure(3);
>> plot(x,y,x,z);
>> legend('sqrt','pow2');
```

Example 2.19: Plot commands for multiple functions in one window

Figure Hold

Multiple arrays with different dimensions can be plotted in the same figure by using `hold on` and `hold off`. The same figure window is held open for each executed `plot` command between `hold on` and `hold off`. (Example 2.20, Figure 2.8)

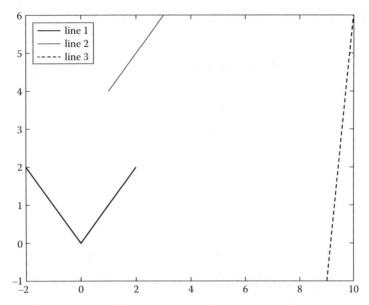

Figure 2.8: Example of hold on and hold off functions

Examples: Execute the following commands in MATLAB.

```
>> x = [ -2 ; -1 ; 0 ; 1 ; 2 ];
>> y = [ abs(-2) ; abs(-1) ; abs(0) ; abs(1) ; abs(2) ];
>> a = [ 1 ; 2 ; 3 ];
>> b = [ 4 ; 5 ; 6 ];
>> plot(x,y);
>> hold on;
>> plot(a,b);
>> plot([9 ; 10], [-1 ; 6]);
>> legend('line1','line2','line3');
>> hold off;
```

Example 2.20: Plot commands hold on and hold off

Subplot

Separate axes (and arrays) can be plotted within the same figure by using the subplot function. This function specifies the number of axes and their arrangement within a single figure. The MATLAB documention for subplot provides additional details about its use. It is demonstrated in Example 2.21 and Figure 2.9.

Examples: Execute the following commands in MATLAB.

```
>> x = [ -2 ; -1 ; 0 ; 1 ; 2 ];
>> y = [ abs(-2) ; abs(-1) ; abs(0) ; abs(1) ; abs(2) ];
>> a = [ 1 ; 2 ; 3 ];
>> b = [ 4 ; 5 ; 6 ];
>> subplot(2,1,1);
>> plot(x,y);
>> subplot(2,1,2);
>> plot(a,b);
```

Example 2.21: Subplot commands

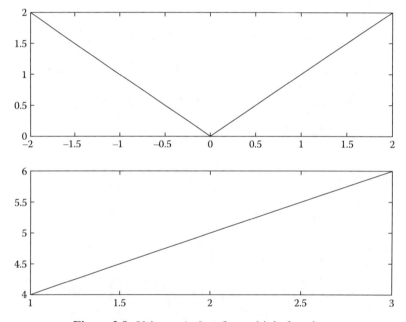

Figure 2.9: Using subplot for multiple functions

Chapter 3

Basics of Audio in MATLAB®
Creating, Importing, and Indexing Arrays

3.1 Introduction: Digital Audio Signals

An audio signal can be represented as a function of amplitude over time: $A(t)$. When the signal is in the form of acoustic sound in physical space, it is a function of sound pressure over time. A microphone converts sound pressure to another form: electricity. In this case, the audio signal is a function of voltage over time or current over time. Sound pressure and electricity are functions of continuous time, meaning an amplitude value exists for all time during a signal.

A computer processes another form of the audio signal, a digital respresentation. The values in a digital signal are based on sampled (i.e., measured) amplitude values of the analog signal. The samples in a digital signal occur at regular time intervals, T_s, called the **sampling period** with units of seconds per sample. The **sampling rate**, $F_s = \frac{1}{T_s}$, of a digital signal defines the number of samples per second. Related to the sampling rate is the **Nyquist frequency**, $\frac{F_s}{2}$. An analog signal can be digitally sampled and then reconstructed back to an analog signal as long as all of the frequencies contained in the signal are less than the Nyquist frequency (Nyquist, 1928; Shannon, 1949).

Digital signals are functions of discrete time, meaning an amplitude value exists only at the time a sample occurs. Digital signals are written, and distinguished from analog signals, using the notation $A[n]$. A signal $x[n]$ has individual samples $x[1], x[2], ..., x[N]$, where N is the total number of samples in the signal. An equivalent notation adopted in this text for a signal is $x = \{x_1, x_2, ..., x_N\}$.

In MATLAB, a digital audio signal can be represented by a sequence of numbers, as an array. Mono audio signals are represented by an array with one column and many horizontal rows representing different samples at regular intervals of time. Stereo audio signals (Chapter 9) are represented by an array with two columns, where the first column is the *left* channel and the second column is the *right* channel. The value in each row corresponds to the amplitude of the audio signal for that sample number.

3.2 MATLAB Audio Functions

There are several functions in MATLAB for working with audio. These functions include `audioread`, `audiowrite`, `audioinfo`, and `sound`.

3.2.1 audioread

The `audioread` function can be used to import an audio file into MATLAB. The samples in the audio file are stored in a column vector. Additionally, the sample rate can be read from the audio file and assigned to a scalar variable. The syntax for using the function is the following: `[y,Fs] = audioread(filename)`.

It is necessary for the variable, `filename`, to be a string with the name of an audio file stored on the computer. Types of supported files for the `audioread` function are .wav, .flac, .mp3, .ogg, .aiff, and several others. The file extension should be included in the `filename`.

In Example 3.1, the syntax for the `audioread` function is demonstrated. The result of the example is an array, `sw`, containing the samples of the signal and a variable, `Fs`, with the sampling rate of the file stored on the computer. The variables are printed to the Command Window to view. Throughout the book, sound files referenced in examples are included in the book's *Additional Content* (www.hackaudio.com).

Examples: Execute the following commands in MATLAB.

```
>> fn = 'sw440Hz.wav';
>> [sw,Fs] = audioread(fn)
```

Example 3.1: Function: `audioread`

3.2.2 sound

The `sound` function can be used to play a MATLAB array of audio samples through the computer's output device. It is necessary to specify the appropriate sampling rate, `Fs`, for playback. The syntax for using the function is the following: `sound(y,Fs)`. The commands to import a sound file and play it are demonstrated in Example 3.2.

Examples: Execute the following commands in MATLAB.

```
>> fn = 'sw440Hz.wav';
>> [sw,Fs] = audioread(fn);
>> sound(sw,Fs);
```

Example 3.2: Function: `sound`

3.2.3 `audiowrite`

The `audiowrite` function can be used to create an audio file from a MATLAB array, which is then stored on the computer. The name of the file is specified by a string, *filename*. Folder location can also be referenced in the name of the file. The name of the MATLAB array and the appropriate sampling rate are specified by the next two variables, respectively. The syntax for using the function is the following: `audiowrite(filename,y,Fs)`. Executing the commands in Example 3.3 creates a new sound file saved to the hard drive of the computer.

Examples: Execute the following commands in MATLAB.

```
>> fnIn = 'sw440Hz.wav';
>> [sw,Fs] = audioread(fnIn);
>> fnOut = 'testSignal.wav';
>> audiowrite(fnOut,sw,Fs);
```

Example 3.3: Function: `audiowrite`

3.2.4 `audioinfo`

The `audioinfo` function can be used to find out information about the contents of an audio file. This information includes the name of the file, whether the file is mono or stereo, sampling rate, number of bits per sample, as well as other information. The syntax for using the function is the following: `info = audioinfo(filename)`.

Examples: Execute the following commands in MATLAB.

```
>> fn = 'sw440Hz.wav';
>> info = audioinfo(fn)
```

Example 3.4: Function: `audioinfo`

3.3 Working with Audio Signals in Arrays

In addition to the built-in functions, there are several foundational methods of working with audio signals stored in arrays. These methods include accessing a portion of a signal, creating new signals, and determining information about the array.

3.3.1 Indexing Arrays

An array contains a set of information. A programmer can access a subset of the information (i.e., the elements) by **indexing** the array, also called **referencing**. Elements in MATLAB arrays can be indexed for both *reading* and *writing* purposes. Vector elements are indexed by a single position number, whereas matrix elements can be indexed by a row number and column number. More information can be found in the MATLAB help documentation for Matrix Indexing.

$$v = \begin{bmatrix} v(1) & v(2) & v(3) & \dots \end{bmatrix}$$

$$v = \begin{bmatrix} v(1) \\ v(2) \\ v(3) \\ \vdots \end{bmatrix}$$

$$m = \begin{bmatrix} m(1,1) & m(1,2) & m(1,3) \\ m(2,1) & m(2,2) & m(2,3) & \dots \\ m(3,1) & m(3,2) & m(3,3) \\ & \vdots \end{bmatrix}$$

A special index can be used for the final element in an array by using the MATLAB keyword end.

Examples: Execute the following commands in MATLAB.

```
>> V = [ 20 21 22 23 24 25 ];
>> V(1), V(3), V(6), V(end)
>> V(2) = 31;
>> V
>> M = [ 11 12 ; 13 14 ; 15 16 ];
>> M(1,end), M(end,2)
>> M(end,end) = 100
>> [sw,Fs] = audioread('sw440Hz.wav');
>> sw(1), sw(2)
```

Example 3.5: Indexing arrays

Elements in arrays can be indexed individually or in groups, allowing multiple elements to be indexed together. A range of elements can be addressed in a vector or a matrix by using a colon. With matrices, using a colon without number indices addresses all elements in the row or column. Using a colon with number indices addresses elements in the specified rows or columns.

Examples: Execute the following commands in MATLAB.

```
>> vv = V(:)
>> vv(:)
>> vv = V(2:3)
>> vv = V(4:end)
>> V(1:2) = [ 41 42 ]
>> mm = M(:,1)
>> mm = M(2,:)
>> mm = M(:,:)
>> M(2,:) = [ 41 42 ]
>> mm = M(2:3,1:2)
>> mm = M(1:2,2)
>> [sw,Fs] = audioread('sw440Hz.wav');
>> plot(sw(1:201),1);
```

Example 3.6: Indexing matrices

Audio Edit Signal Splice

One way to edit an audio signal is to perform a **splice**. This process with digital signals re-creates the editing process of putting a physical splice into analog magnetic tape. When a signal is recorded on tape, an audio engineer can separate the signal by cutting the tape into two or more parts. For digital signals, this is accomplished by indexing all the samples up to the sample number the splice occurs.

The opposite of signal splicing is putting two or more signals together. With analog tape, an audio engineer could use an adhesive to hold two separate pieces of tape together, effectively repairing the splice. For digital signals, separate segments of a signal can be combined into a new signal by using array concatenation (Section 2.6). A demonstration of editing a vocal recording to remove a segment is shown in Example 3.7.

Examples: Execute the following commands in MATLAB.

```
>> [x,Fs] = audioread('Vocal.wav');
>> plot(x);
>> edit1 = x(1:47609,1);
>> plot(edit1,1);
>> sound(edit1,Fs);
>> edit2 = x(123252:end,1);
>> sound(edit2,Fs);
```

```
>> edit3 = [edit1 ; edit2];
>> sound(edit3,Fs);
```

Example 3.7: Signal splice

3.3.2 *Array Reversal*

A simple method to change an array is to reverse the order of the sequence of samples. For an audio signal, this performs a time reversal, creating a backwards signal. It is accomplished by indexing the elements in an audio signal array starting at the end and continuing until the first element.

Examples: Execute the following commands in MATLAB.

```
>> [x,Fs] = audioread('Vocal.wav');
>> xReverse = x(end:-1:1);
>> sound(xReverse,Fs);
```

Example 3.8: Time-reversing a vocal recording

3.3.3 *Additional Methods to Create Arrays*

MATLAB provides several methods for efficient programming to create arrays. A row vector of increasing integers can be created by specifying the first term and last term, separated by a colon. A row vector with constant spacing can be created by specifying the first term, the spacing, and the last term, separated by colons.

Examples: Execute the following commands in MATLAB.

```
>> vec1 = 6:10
>> vec2 = [ -4:3 ]
>> vec3 = 1:2:11
>> vec4 = 1:2:10
>> vec5 = 1:3:2
>> vecErr = [ 3:1 ]
>> vec6 = 3:-1:1
>> Ts = 1/48000;
>> t = 0:Ts:1
```

Example 3.9: Efficient array creation

A row vector with linearly spaced values can be created by specifying the first term, the last term, and the number of terms, using the `linspace` function. MATLAB determines the correct spacing based on the provided specifications. The linspace function is called by entering linspace(firstElement, lastElement, numberOfElements). *Note:* The number of elements must be a positive, whole number.

Examples: Execute the following commands in MATLAB.

```
>> vec7 = linspace(10,14,5)
>> vec8 = linspace(10,13,5)
>> vec9 = linspace(13,10,5)
>> vec10 = linspace(0,100,5)
>> mat5 = [ 1:2:11 ; linspace(1,11,6) ]
```

Example 3.10: Function: linspace

A row vector with logarithmic spacing can be created using a similar syntax with the logspace function. MATLAB determines the correct spacing based on the provided specifications. The logspace function is called by entering logspace(a,b, numberofElements), creating an array of values between 10^a to 10^b. *Note:* The number of elements must be a positive, whole number.

Examples: Execute the following commands in MATLAB.

```
>> vec11 = logspace(0,2,10)
>> vec12 = logspace(1,4,20)
```

Example 3.11: Function: logspace

There are several functions in MATLAB to create arrays with predefined values. Examples are ones(r,c), zeros(r,c), and eye(d). The function ones(r,c) creates an array with all values set to 1. The function zeros(r,c) creates an array with all values set to 0. The number of rows is given by the input value, r, and the number of columns is given by the input value, c. The function eye(d) creates an *identity* matrix with diagonal values set to 1, and all remaining values set to 0. The number of rows and columns is identical in the identity matrix, given by the input value d.

Examples: Execute the following commands in MATLAB.

```
>> ones(3,1)
>> zeros(2,3)
>> eye(4)
```

Example 3.12: Arrays with predefined values

3.3.4 Array Transposition

Transposition is an array operation such that a column vector becomes a row vector or vice versa. Matrices can also be transposed to interchange columns and rows. The **transpose operator** is a period and comment symbol (.'). The colon operator (:) can be used to turn any array into a column vector.

Examples: Execute the following commands in MATLAB.

```
>> x = [ 1 2 3 ];
>> traVec = x.'
>> w = [ 1 2 3 ; 4 5 6 ];
>> traMat = w.'
>> colVec = x(:)
>> colVec(:)
>> w(:)
```

Example 3.13: Array transpose operation

3.3.5 Determining Dimensions of Arrays

MATLAB has several functions to determine the dimensions of vectors and matrices. The function size(x) determines the row and column dimensions of array, x. Separate values for the row and column dimensions can be assigned from the function using the command [row,col] = size(x). Using the function size(x,1) returns only the number of rows and size(x,2) returns only the number of columns. The function length(x) determines the maximum dimension (either row or column) of array x.

Examples: Execute the following commands in MATLAB.

```
>> x = ones(3,2)
>> [row,col] = size(x)
>> size(x,1)
>> size(x,2)
>> length(x)
>> y = zeros(2,3)
>> length(y)
```

Example 3.14: Determining dimensions of arrays

3.4 *Visualizing the Waveform of an Audio Signal*

An array containing an audio signal can be plotted similar to other arrays in MATLAB. If an array of an audio signal is used as the only input variable to the plot function, the figure displays the signal's amplitude versus the sample number. To display the signal's amplitude versus time in seconds, it is necessary to include an additional array as an input variable to the plot function, which contains the time in seconds each corresponding sample occurs. The visualization of a signal's amplitude versus time is called a **waveform**.

Examples: Execute the following commands in MATLAB.

```
>> [sw,Fs] = audioread('sw20Hz.wav');
>> plot(sw); xlabel('Sample Number');
>> Ts = 1/Fs;
>> t = 0:Ts:1;
>> figure;
>> plot(t,sw); xlabel('Time (sec.)');
```

Example 3.15: Plotting audio signals

The output of each plot function is shown in Figures 3.1 (amplitude of the signal versus sample number) and 3.2 (amplitude of the signal versus time).

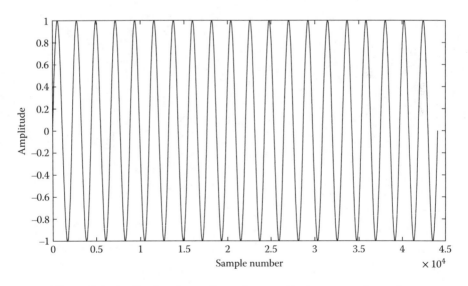

Figure 3.1: Audio signal waveform plot: amplitude vs. sample number

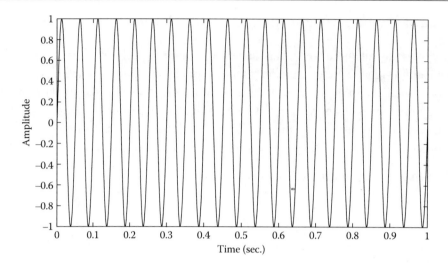

Figure 3.2: Audio signal waveform plot: amplitude vs. time

Bibliography

Nyquist, H. (1928). Certain topics in telegraph transmission theory. *Transactions of the American Institute of Electrical Engineers, 47*(2), 617–644. doi:10.1109/t-aiee.1928.5055024

Shannon, C. E. (1949). Communication in the presence of noise. *Proceedings of the Institute of Radio Engineers, 37*(1), 10–21. doi:10.1109/jrproc.1949.232969

Chapter 4

MATLAB® Programming Environment
Command Window, Editor, m-Files, and Documentation

4.1 Introduction: MATLAB Application

In order to program more complicated things with audio signals, it is necessary to cover additional information about MATLAB and programming in general. This chapter discusses these topics by building on the foundation of the basics in Chapters 2 and 3.

MATLAB is a computer programming language and also an application that can be run on a personal computer. The software application is compatible with Windows, Mac, and Linux.

The MATLAB application is a **programming environment** comprised of multiple windows and menus. The shape and size of the windows and menus is customizable. Several windows and menus can be either hidden or displayed, docked or undocked, based on user preference. More information can be found on the MathWorks website under the *Getting Started* tutorials.

In addition to MATLAB, there are add-on toolboxes that include extra specialized features for the software. Throughout this book, features of the Signal Processing Toolbox are demonstrated and used.

4.1.1 Command Window

The MATLAB **Command Window** can be used for executing commands, running programs, opening other windows and files, and managing the software. Commands are written and executed by typing next to the command prompt (≫). Commands that were previously executed can be recalled by pressing the up-arrow key (↑) and down-arrow key (↓). These commands can be modified and executed, if needed. The text in the Command Window can be cleared by entering the clc command. Previously executed commands remain in the **Command History** Window.

Execute Multiple Commands

MATLAB can execute multiple commands in sequential order from the Command Window. The order of execution is from left to right and top to bottom.

Multiple assignment statements can be included on the same command line if they are separated by a comma. Additionally, multiple assignments statements can be included on the same line (and the output surpressed) if they are separated by a semicolon. The order of execution is from left to right on the same line.

Examples: Execute the following commands in MATLAB.

```
>> start = 2.5, stop = 5.0
>> a = 3; b = 4, c = 5; d = 6
>> q = 1, w = q
>> gErr = k, k = 1
```

Example 4.1: Executing multiple commands on the same line

There are several methods to execute commands on multiple lines. Rather than only pressing the [RETURN] key after a command is entered, if the combination of [SHIFT] and [RETURN] keys is pressed after a command, this command is not executed until the [RETURN] key is pressed on its own. Another method to execute mutliple commands is to add three dots, . . . , at the end of a single command.

Examples: Execute the following commands in MATLAB.

```
>> width = 2.5 [SHIFT] [RETURN]
height = 3.75
>> width = 4 ... [RETURN]
height = 6
>> width = 5, ... [RETURN]
height = 7; ... [RETURN]
depth = 8
>> score1 = 40, ...
score2 = score1
>> score3 = score4, ...
score4 = 20
```

Example 4.2: Execute multiple commands

4.1.2 Workspace

The variables created and stored in computer memory can be viewed and accessed from the **Workspace** browser. Variables displayed in the workspace can be viewed, modified, and plotted.

Clear the Workspace

The variables in the MATLAB workspace (and computer memory) can be removed or cleared using the `clear` command. This command removes all variables from memory. Individual variables can be removed from memory by typing the command `clear x price timeRemaining`, as an example.

Examples: Execute the following commands in MATLAB.

```
>> x = 5;
>> price = 3;
>> timeRemaining = '20 seconds';
>> clear x
>> clear price timeRemaining
```

Example 4.3: Clear the MATLAB workspace

Save and Load Variables in the Workspace

The variables in the workspace can be saved and loaded as a MATLAB ".mat" file. The variables can be saved by using the drop-down arrow in the Workspace Window. Additionally, the command `save filename` can be entered in the Command Window. Individual variables can be saved by following the command with the names of the desired variables.

Variables can be loaded into the workspace by using the *Open* button or the *Import Data* button in the MATLAB Toolbar. Additionally, the command `load filename` can be entered in the Command Window.

Examples: Execute the following commands in MATLAB.

```
>> t = [ 5 6 7];
>> x = 8;
>> save test1
>> save test2 x
>> clear
>> load test1
```

```
>> clear
>> load test2
```

Example 4.4: Saving and loading in the MATLAB workspace

4.1.3 Current Folder

When MATLAB is accessing files stored on the computer, it references these files from the **current folder**. The filepath of the current folder can be found by entering the command pwd. This folder can be changed by using the Current Folder menu. Additionally, the current folder can be changed by entering the command cd('folderName').

A new folder can be created by using the function mkdir('folderName'). A folder can be removed by using the function rmdir('folderName').

Examples: Execute the following commands in MATLAB.

```
>> oldDir = pwd;
>> newDir = 'SubFolder1';
>> mkdir(newDir);
>> cd(newDir)
>> pwd
>> cd(oldDir)
>> pwd
>> ls
>> rmdir(newDir)
>> ls
```

Example 4.5: MATLAB current folder

4.2 MATLAB m-Files

A MATLAB **m-file** (.m) is a file that can contain commands to be executed in MATLAB. This type of file is one way commands can be grouped together into a computer program. An m-file can be created, saved, loaded, edited, and executed. Using these files can be helpful if there are multiple commands to be executed. There are several different types of m-files including scripts and functions (Section 5.3.3). Example 4.6 shows an m-file with multiple commands written together.

Examples: Save and run the following m-file in MATLAB.

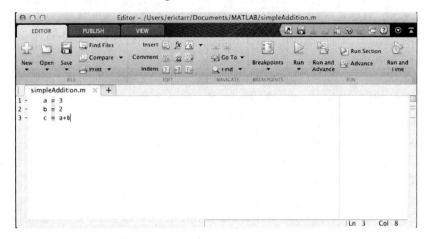

Example 4.6: MATLAB m-file (`simpleAddition.m`). (Reprinted with permission of The MathWorks, Inc.)

4.2.1 Scripts

A **script** file is a sequence of commands to be executed together. A script can be run by clicking the *Run* button on the menu of the Editor Window. Additionally, a script can be run by entering the name of the script in the Command Window. When the Editor is selected, the keyboard shortcut [option]-[command]-[R] can also be used to run a script from the Mac OS application. When a script file is run, the commands are executed from top to bottom unless a programmer changes the flow of execution (Chapter 5). The values of created variables are saved to the MATLAB workspace. The output of each command that generates output is displayed in the Command Window.

Examples: Save and run the following m-file in MATLAB.

```
filename = 'Vocal.wav';
[sig,Fs] = audioread(filename);

figure(1);
plot(sig);
sound(sig,Fs);
rev = sig(end:-1:1);

figure(2);
```

```
plot(rev);
sound(rev,Fs);
```

Example 4.7: Script: `signalReverse.m`

4.2.2 Commenting Code

Text to the right of a percent symbol, %, is called a **comment**. When MATLAB is executing the code in an m-file, it does not execute comments. This aspect of a programming language allows a programmer to include information, such as documention, in a script that does not follow the syntax necessary to be executed. More generally, it is a good practice to add comments to a computer program to describe the purpose of the code. In some cases, it is helpful to provide comments for individual commands. It can also be appropriate to include comments to describe the execution of groups of commands that perform a larger task.

A comment can be written on its own line in a MATLAB m-file. It can also be written to the right of an executed command when separated by a percent symbol, %.

Examples: Save and run the following m-file in MATLAB.

```
% This script calculates the square root of three numbers.

in1 = 36; % Assign the number 36 to the variable "in1"
in2 = 144;
in3 = 16;

% Call the function 'sqrt' to calculate the square root
% of each variable

out1 = sqrt(in1)

out2 = sqrt(in2)     % The result is printed to the Command Window

out3 = sqrt(in3)
```

Example 4.8: Adding comments to a MATLAB m-file

4.3 MATLAB Debugging Mode

The MATLAB programming environment includes a mode for troubleshooting or debugging errors in a program. This can be useful when trying to understand how a program works, and also when searching for problems to fix. MATLAB files can be debugged using the Editor Window, as well as using debugging functions from the Command Window (e.g., dbcont, dbstep, dbquit, etc.).

Examples: Run the following m-file in MATLAB using the specified breakpoints.

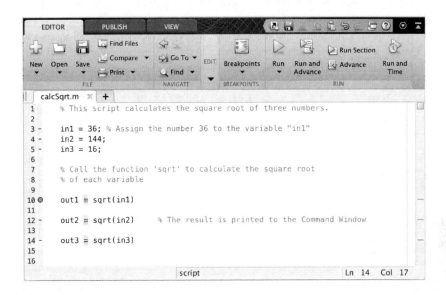

Example 4.9: MATLAB debugging breakpoint example. (Reprinted with permission of The MathWorks, Inc.)

The primary method for debugging a MATLAB program is to set a breakpoint. When a program is run, all commands are executed prior to the line with the first breakpoint. Then the program stops before executing the command on the line with the breakpoint. This allows the programmer to analyze the program in its current state, including variables and their values. In Example 4.9, a script is shown with a breakpoint inserted on line 10 by clicking the dash next to the line number.

When the program is paused, it can be resumed in several different ways. Pressing the *Continue* button in the Editor Window starts executing commands until the next breakpoint is reached or the program finishes. Pressing the *Step* button executes the current line of code in the current file, then pauses at the next line. Pressing the *Step In* button executes the current line of code and continues debug mode within any functions that are called.

It is necessary to quit debug mode to edit the code of the computer program. To run the program without entering debug mode, it is necessary to remove any inserted breakpoints. This is accomplished by pressing the *Clear All Breakpoints* button from the Editor Window, or by executing the function dbclear in the Command Window.

4.4 MATLAB Help Documentation

There is extensive documentation available for many aspects of MATLAB including Getting Started, Functions, and so on. This can be a helpful resource to refer to while learning the software and working with the built-in functions. This information is available using the Help menu or the Search Documentation toolbar. Examples of the MATLAB documentation are shown in Figures 4.1 and 4.2. The same documentation can also be accessed by entering the command help in the Command Window. Information for specific functions can be accessed by entering the command help function.

Examples: Execute the following commands in MATLAB.

```
>> help
>> help round
>> help ceil
>> help linspace
>> help dec2bin
```

Example 4.10: MATLAB help commands

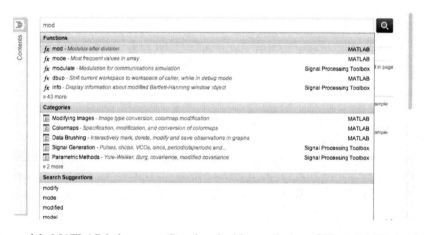

Figure 4.1: MATLAB help menu. (Reprinted with permission of The MathWorks, Inc.)

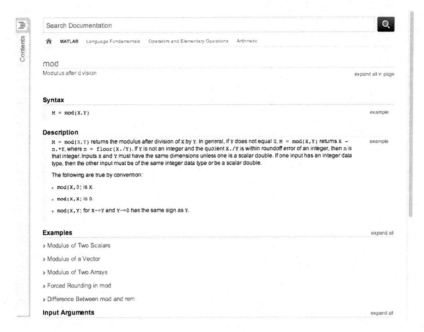

Figure 4.2: MATLAB help documentation. (Reprinted with permission of The MathWorks, Inc.)

Chapter 5

Logicals and Control Structures in Programming
Conditional Statements, Loops, and Functions

5.1 Introduction: Controlling the Flow of Execution

When a computer program is run, the commands are executed in a particular order. As discussed in Chapter 4, the order is usually top to bottom, left to right for the commands in a script. This does not always need to be the case, as a programmer can instruct the computer to follow a different order.

Many programming languages, including MATLAB, include designated keywords to change the order of executed commands. These parts of a programming language are called control structures. Additional documentation related to this chapter can be found on the MathWorks website under *Control Flow*.

For the various control structures, the MATLAB programming language makes use of a special data type called a logical. In this chapter, background information is introduced about logicals first, and then it is put to use with control structures.

5.2 Logical Data Type

Logicals are a different data type than numbers (Section 2.5.1) and text (Section 2.5.2). They are a data type that can only have a value of either TRUE or FALSE. Because there are only two possible values, MATLAB designates logicals as a specific data type. Values of logicals can be 1, 0, `true`, or `false`. Other names used in computer programming for a logical are a boolean or a bool. These names are a reference to the mathematician George Boole, who developed the algebraic system of logic used in computer programming.

5.2.1 Logical Operations

Besides executing mathematical operations, MATLAB (and other programming languages) can execute **logical operations**. A logical operation can be used to determine if an expression is TRUE or FALSE. By convention, when MATLAB executes a logical expression, the result is the value 1 when TRUE and the value 0 when FALSE.

There are several logical operations available in MATLAB. Operator symbols and descriptions are shown in Table 5.1. More information can be found in the MATLAB help documentation for *Relational Operators*.

The "equal to" logical operation consists of two equal signs (==), because the single equal sign (=) is used for the assignment operator. The "not equal to" logical operation consists of the tilde symbol followed by an equal sign (~=).

Examples: Execute the following commands in MATLAB.

```
>> 5 == 5    % This is true, observe value of "ans"
>> 4 == 5    % False, observe new value of "ans"
>> 4 < 5
>> 4 > 5
>> m = 5     % This assigns value of 5 to "m"
>> m == 5    % This tests whether value of m is equal to 5
>> m == 4
>> m ~= 5    % "Not equal to" operator, opposite of equal
>> m == 4 + 1    % Logical operators can be combined with math
>> false == 0    % Keywords false and true can be compared to values
```

Example 5.1: Logical operations

Just as the result of mathematical operations can be assigned to a variable, the result (0 or 1) of logical expressions can be assigned to a variable. The result (0 or 1) of logical expressions can be assigned to a variable (Example 5.2).

Table 5.1: Logical operators in MATLAB

Operator	Description
<	Less than
<=	Less than or equal to
>	Greater than
>=	Greater than or equal to
==	Equal to
~=	Not equal to

Examples: Execute the following commands in MATLAB.

```
>> d = 1 == 1     % Result of comparison (1==1) stored in variable "d"
>> d == true      % Use variable "d" for other comparisons
>> d == false
>> jj = 13;       % These statements can get complicated
>> test1 = jj == 4 * 3   % Start from right and work backwards
>> test2 = test1 == 1    % First perform comparison, then assignment
```

Example 5.2: Assigning logical operations to variables

5.2.2 Combining Logical Operators

Multiple logical operators can be combined to form more complex logical statements using the & (AND) and | (OR) operators. The result of multiple logical statements combined with the & (AND) operator are always FALSE, unless all individual logical statements are TRUE. The result of multiple logical statements combined with the | (OR) operator are always TRUE, unless all individual logical statements are FALSE.

Examples: Execute the following commands in MATLAB.

```
>> 5 == 5 & 4 == 4    % Observe result of "ans" for various
>> 5 == 5 & 4 == 3    % combinations of TRUE, FALSE, AND, OR
>> 5 == 5 | 4 == 3
>> 5 == 4 & 4 == 3
>> 5 == 4 | 4 == 3
```

Example 5.3: Combining logical operations

Multiple logical operators can also be combined using the and and or functions. More information can be found in the MATLAB help documentation for *Logical Operators*.

Examples: Execute the following commands in MATLAB.

```
>> A = 1; B = 1; C = 0; D = 0;
>> and(A,B)    % Observe result of "ans" for various
>> or(A,B)     % examples using MATLAB logical functions
>> and(A,C)
>> or(A,C)
>> and(C,D)
>> or(C,D)
```

Example 5.4: MATLAB logical functions

5.2.3 String Compare

A different logical operation can be used specifically for comparing entire strings. By default, the "equal to" operation performs an element-wise comparison between strings. In other words, each individual character is compared with the corresponding character in the other string. A logical value is returned for each element of the string.

The MATLAB function `strcmp` can be used to compare entire strings. In this case, a single logical is returned for the entire operation. If all the individual characters are equal, the result is TRUE. If one or more characters are not equal, the result is FALSE.

Examples: Execute the following commands in MATLAB.

```
>> 1 == '1'        % Observe if a number equals a character
>> '1' == '1'      % Observe if == can be used for characters
>> 'Q1!' == 'Q1!'  % Observe result if == used for strings
>> FF = 'Q1!'; GG = 'Q1!'; HH = 'q1!';
>> FF == GG        % Observe result of string comparison if
>> FF == HH        % only difference is capitalization
>> strcmp(FF,GG)   % Use strmp to test if entire strings match
>> strcmp(FF,HH)
```

Example 5.5: String compare

Conceptually, logical operations allow a computer to ask and answer a TRUE/FALSE question while executing code. This ability to resolve a TRUE/FALSE question makes it possible to use logical operations as part of more sophisticated tasks, such as control structures.

5.3 Types of Control Structures

There are several different types of **control structures**. Each type instructs the computer to change the flow of executed commands in different ways. Control structures can be used to execute command statements under certain conditions, repeat the same command statements multiple times, and execute commands located in other files.

5.3.1 Conditional Statements

A conditional statement is a command that makes a decision whether or not to execute the group of commands based on a condition. Conditional statements use a logical expression to determine which commands to execute. If the result of the logical is TRUE, a set of commands is executed. If FALSE, the set of commands is not executed. More information can be found in the MATLAB help documentation for *Conditional Statements*.

if *Statement*

An if **statement** executes a logical expression to decide whether or not to execute the commands following the conditional statement down to the accompanying end statement. If the logical expression is TRUE the commands are executed. Otherwise, if it is FALSE, the commands are not executed and the flow of execution continues after the end statement.

Examples: Save and run the following m-file in MATLAB.

```
% This script demonstrates several examples of "if" statements.
% Observe which commands get executed, and which do not, based
% on the conditional statements

TT = 2;        % Variable for comparison
if TT == 2     % This is true
    SS = 3     % Therefore, this is executed
end

if TT == 0     % This is false
    RR = 3     % Therefore, this is NOT executed
end            % Flow of execution skips to 'end'

if TT > 1      % This is true
    QQ = 3
end

if TT < 1      % This is false
    PP = 3
end
```

Example 5.6: if conditional statement

else *and* elseif *Statements*

Including else and elseif **statements** within an if conditional statement allows for another group of commands to be excuted when the logical expression of the if statement is FALSE. These statements cannot be used unless they are between an if statement and the accompanying end statement.

An else statement is only executed when all other conditional statements (if and elseif) are FALSE. An else statement does not have a logical expression explicitly associated with it; it is implicitly assumed all other conditional statements are FALSE. There can only be one else statement within an if statement.

One or more elseif statements can be included within an if statement as an additional conditional statement. An elseif statement requires a logical expression. When the logical expression is TRUE, the commands following the elseif statement and before the next conditional statement are executed. Otherwise, if it is FALSE, the commands are not executed. There can be multiple elseif statements included within an if statement.

Examples: Save and run the following m-file in MATLAB.

```matlab
% This script demonstrates several examples of "elseif"
% and "else" statements. Observe which commands get executed,
% and which do not, based on the conditional statements


num = -1;          % Variable for comparison
if num > 0         % This is false
    gtz = true     % Therefore this is NOT executed
else               % Instead,
    gtz = false    % This is executed
end



num = -1;
if num > 0         % This is false
    gtz = true
elseif num < 0     % This is true
    ltz = true     % Execute this command
else               % Skip all other commands, jump to end
    zero = true
end



num = 0;
if num > 0         % This is false
    gtz = true
elseif num < 0     % And this is false
    ltz = true
else               % Therefore,
    zero = true    % Only execute this command
end
```

Example 5.7: else and elseif conditional statements

`switch` *Statement*

A `switch` **statement** conditionally executes a set of statements contained in several cases. Similar to an `if` statement, only certain commands are executed during the program depending on the conditions of the `switch` statement. A `switch` statement evaluates whether a variable is equal to the variable in each case. If it is TRUE, the statements in that case are executed. Otherwise, the program moves on to testing the next case.

Examples: Save and run the following m-file in MATLAB.

```
% This script is an example of using a "switch" statement
% as a conditional statement

a = 5;          % Variable for switch cases
switch a
    case -5    % "a" does not equal -5
        b = 1  % Skip this case

    case 5     % "a" does equal 5
        b = 2  % Execute this case
end
```

Example 5.8: `switch` conditional statement

Comparison of `if` *and* `switch` *Statements*

The `if` and `switch` conditional statements can be used to accomplish identical flow in a computer program. Therefore, in most cases it is the preference of the programmer to decide which type of statement to use. As a matter of best practice, a programmer should use the conditional statement that allows for the easiest interpretation of the written code.

5.3.2 Loops

One task computers excel at performing is the repetition of instructions. As a programmer, if you want to repeat a command multiple times, you could copy and paste the command manually. Another option is to create a **loop**. A loop repeats the included commands until certain requirements are met. Requirements are defined in the command used to create the loop, also called the loop declaration.

`for` *Statement*

One type of loop is declared in MATLAB using the `for` **statement**. Following the `for` keyword, an expression is needed to define a counting variable to set the number of times the loop repeats. This variable can be referenced during the repetitions of the loop. The `for` loop needs to be concluded by the keyword `end` to indicate which commands are repeated.

Examples: Save and run the following m-file in MATLAB.

```
% This script demonstrates several examples of "for" loops.
% Observe how the commands within the loop are executed
% and how the update variable changes.
% Recommendation: set breakpoints before a loop. During
% debug mode, use "step" and observe the Workspace to
% observe how variables change.

% Declare the beginning of the "loop"
for ii = 1     % The counting variable gets the value "1"
    ii         % Print the value of the counting variable
end


% Declare another loop
for ii = 1:3   % Set the variable to count from 1 to 3
    ii         % Print the value each time in the loop
end


% The counting variable in this loop goes from 1 to 7
for ii = 1:2:7
    ii      % A step size of 2 is used for each repetition
end


% This loop calculates various values of the sine function
for ii = 0:pi/2:2*pi   % {0, pi/2, pi, 3*pi/2, 2*pi}
    % Use the value of the variable in a mathematical expression
    sin(ii)
end
```

Example 5.9: Script: `forLoopExample1.m`

The variable declared with the `for` loop can be used for counting, as part of mathematical expressions, and as an index for arrays. This is a useful approach to indexing each element in an array, one by one, from start to finish.

Examples: Save and run the following m-file in MATLAB.

```
% This script demonstrates several examples of using
% the iteration variable for different purposes within
```

```
% a loop.


% Example array for indexing
a = [ -0.2, 0.4, -0.8, 0.5, 0.1, -0.6];


for element = 1:6  % Declare loop to index each element
    a(element)     % Print an individual element
end

% Use looping variable for creating a new array
for t = 1:5
    h(t) = 0;
end
h               % Print the array after loop is complete


% Use variable with both an array and mathematical expression


for t = 1:5
    hh(t) = t+100  % Conceptually, "hh(t)" is a mathematical function
                   % over the variable "t" with values {1,2,3,4,5}

end


Ts = (1/48000);    % Assume a "sampling period" Ts
for n = 1:10
    % Relationship between sample number and time in seconds
    t(n) = n*Ts;
end


sampleNum = 1;     % Initialize additional counting variable
% This loop uses a variable "t" with a fractional step size
for t = 0:0.1:1
    % Value of "t" is assigned to array at integer index
    k(sampleNum) = t
    sampleNum = sampleNum + 1 % Iterate integer variable
end
```

Example 5.10: Script: `forLoopExample2.m`

`while` *Statement*

Another type of loop is declared in MATLAB using a `while` **statement**. Following the `while` keyword, a logical expression is used to define the condition for which the loop executes.

Therefore, a while loop is like a conditional statement (Section 5.3.1) which also repeats. The while loop needs to be concluded by the keyword end to indicate the lines of code to be repeated. Only the commands on lines between while and end are repeated for the loop.

When execution reaches the declaration of the loop, if the logical operation is TRUE, then the code in the loop is executed. When the commands within the loop are complete, execution returns to the logical expression at the loop declaration, and it is evaluated again. As long as the logical expression remains TRUE, the loop continues to repeat. At any point, if the logical expression produces a FALSE result, then execution jumps to the end of the loop.

The following is a conventional approach for using a while loop. For an existing variable in code, the loop is declared using the while keyword followed by a logical expression containing the variable. Any commands to be repeated in a given task are executed within the loop. Also within the loop, the variable used in the original logical expression is changed in some way (e.g., iterative counting), such that the logical expression can eventually change from TRUE to FALSE, so the loop exits. Several examples of this approach are demonstrated in Example 5.11.

Examples: Save and run the following m-file in MATLAB.

```
% This script demonstrates several examples of "while" loops.
% Observe how often the commands within each loop are executed.
% Recommendation: set breakpoints before a loop. During
% debug mode. use "step" and observe the Workspace to
% observe how variables change.


% Initialize a variable to use in the logical expression of the loop
XX = 2;
while XX < 1   % The logical expression is false
    YY = 6     % Therefore. this command is not executed
end



AA = 2;
while AA > 1   % This expression is initially true
    BB = 3     % Therefore. this command is executed
    AA = 0;    % The value of the variable is changed
               % such that the next time the logical
               % expression is evaluted. it is false
               % and the loop with exit
end


CC = 2
```

```
while CC > 1
    CC = 0;    % Notice here, all the commands in the loop
    DD = 3     % are executed even if CC is not > 1
               % Then the logical expression is evaluated
               % and the loop exits
end


EE = 1             % Declare a counting variable
while EE < 4
    FF = 8         % Print this variable each time through
    EE = EE + 1    % Iteratively increase the variable
                   % Eventually the loop exits
end
```

Example 5.11: Script: `whileLoopExample.m`

Indefinite Loop

It is possible for a `while` statement to create a loop that repeats indefinitely. This is typically not the intended use of the statement because it prevents subsequent commands from being executed, thus preventing the script from ever completing. An indefinite (also called an "infinite") loop occurs when the conditional statement never returns a FALSE result after the loop repeats. This type of error can occur if a programmer does not change the variable of the conditional statement within the loop.

Examples: Save and run the following m-file in MATLAB.

```
% This script demonstrates how a loop can continue indefinitely
% if the conditional statement is always true. To stop the loop,
% press [CONTROL] + c during execution.
BB = 2;
while BB > 1 % This loop does not end
    CC = 3     % because BB is always 2
end
```

Example 5.12: Script: `indefiniteLoop.m`

To stop the execution of an indefinite loop, press [CONTROL] + c during the execution of a script. This shortcut only works after switching the focus of the MATLAB application to the Command Window.

Comparison of `for` and `while` Statements

The `while` and `for` loop statements can be used to accomplish identical flow in a computer program. As a matter of best practice, a programmer should use the loop statement that allows for the easiest interpretation of the written code. When the number of repetitions in the loop can be explicitly set prior to the loop, it is often preferred to use the `for` statement. When the number of repetitions cannot be known, it may only be possible to use the `while` loop.

Examples: Create and run the following m-file in MATLAB.

```
% This script compares the process of using a for loop
% with the process of using a while loop. The same
% result can be accomplished with each type of loop by using
% the appropriate syntax.

% for loop example
N = 4;                  % Number of elements for signal
sig1 = zeros(N,1);    % Initialize an output signal
for ii = 1:N
    sig1(ii) = sin(ii*pi/2);  % Fill the output array
                              % with values from the
end                            % sine function
sig1                    % Display result on Command Window

% while loop example
sig2 = zeros(N,1);    % Initialize new output signal
jj = 1;                % Declare a counting variable
while jj ≤ N
    sig2(jj) = sin(jj*pi/2);  % Fill the output array
    jj = jj + 1;                % Iterate counting variable
end
sig2                    % Display result on Command Window
```

Example 5.13: Script: `simpleLoop.m`

In audio signal processing, `for` loops are a common way to iteratively step through each sample of a signal. For many signals in MATLAB, the number of samples can usually be determined. Then, the loop can be set to index through the individual samples or segments of the entire signal.

5.3.3 Functions

Functions are a set of commands grouped together and executed each instance the function is called. MATLAB has many built-in functions including `sqrt`, `sin`, `cos`, `ceil`, and so on.

Functions may have input and output variables, but they are only necessary if required by the particular function. Input variables to functions are enclosed in parantheses after the function name. Multiple input variables are separated by commas. Output variables from functions are enclosed in square brackets to the left of an equal sign and function name. Multiple output variables are separated by commas.

Examples: Execute the following functions in MATLAB.

```
>> x = 9;
>> [y] = sqrt(x)
>> [z] = rem(22,x)
>> [sw,Fs] = audioread('sw440Hz.wav');
>> sound(sw,Fs);
```

Example 5.14: MATLAB basic built-in functions

When a function is called, the input and output variables are passed to and from the function. The output variables are the only variables created and stored in the current MATLAB workspace. A separate workspace is created specifically for the function, and any variables local to the function (created in and only used by the function) are stored on the separate workspace. After a function is completed, the separate workspace for the function is cleared.

Functions can also be created and written in a MATLAB m-file by following the necessary scripting constructs and syntax. The MATLAB keyword `function` should be written on a line before any other executed commands in the m-file. This designates the m-file to be a function instead of a script. Then the function should be written in the form of how it is called, including the function name, as well as input and output variables. The name of the function should be identical to the name of the m-file. The input and output variable names for the function's script file are defined on line 1. Any time the input and output variables are referenced within this function's script, these variable names should be used. However, it is not necessary to use these exact variable names when the function is called. In order to call the created function, it should be located in the current folder of MATLAB.

Examples: Save the following function as an m-file in MATLAB.

```
function [y] = userSqrt(x)    % Function declaration
% Calculate the square-root of the input variable
y = (x)^(1/2);  % Store result in the output variable
```

Example 5.15: Writing help documentation in a function

Examples: Execute the following commands in MATLAB.

```
% These commands demonstrate how to use your own function
>> inVar = 64;
>> [outVar] = userSqrt(inVar);
```

Example 5.16: Using the `userSqrt` function

Functions are an aspect of computer programming that allow code to be reusable and easier to understand. As an example, if there is a set of commands in a script used multiple times, it may be appropriate to put the set of commands in a function. Then the function can be substituted rather than having to put the same set of commands in a script multiple times.

Examples: Create and run the following m-file in MATLAB.

```
in1 = 36;                % Declare a few variables
in2 = 144;               % with different values
in3 = 16;


% Calculate the square-root of the variables
out1 = userSqrt(in1)     % Call the userSqrt function


out2 = userSqrt(in2)     % Print the result to demonstrate
                         % the function works for various
out3 = userSqrt(in3)     % input values
```

Example 5.17: Script: `userSqrtExample.m`

Local Functions

A local function is a special type of function located within the m-file of another function or script. A local function cannot be called and executed except from within the same m-file. The syntax for creating a local function is identical to creating a general function. Multiple local functions can be included within the same m-file.

A local function can be used when a programmer wants to write a function, but does not want it to be able to be called from any other external script or function.

Examples: Save the following function as an m-file in MATLAB.

```
% The function at the top, "mystats",
% can be called from other scripts as long as
% this m-file is also named "mystats"
```

```
function [avg, var] = mystats(x)
N = length(x);
avg = mymean(x,N);       % Call other local functions
var = myvar(x,N,avg);    % from within another function
end


% Functions below are local, and can only be
% called from within this m-file
function a = mymean(u,N)
% Calculate mean: arithmetic average
a = sum(u)/N;
end


function b = myvar(v,N,avg) % Declare another local function
% Calculate variance
b = sum(((v-avg).^2))/N;
end
```

Example 5.18: Declaring local functions

Programmer-Created Documentation

In addition to the built-in MATLAB help documentation discussed in Section 4.4, help documentation can be created for functions written by a programmer. Help documentation provides an explanation of how a function works. It also provides a link to other related functions.

To include help documentation in a function file, the code formatting should be followed. The documentation can be written at the top of the function m-file, or immediately following the function declaration. Each line of the documentation should start with a percent symbol, %. Any function names referenced in the documentation should be written in capital letters. It is good practice to include information about input and output variables, such as constraints on the variable type.

Examples: Save the following function as an m-file in MATLAB.

```
% USERSQRT Calculate the square root.
%     y = userSqrt(x) calculates the square root of 'x'.
%
%     x : scalar input variable
%     y : scalar output variable
%
```

```
%    See also SQRT

function [y] = userSqrt(x)

y = (x)^(1/2);
```

Example 5.19: Function: `userSqrt.m`

User-created help documentation can be displayed by entering the command `help functionname` in the Command Window. As an example, entering the command `help userSqrt` displays the documentation included in the function file.

Chapter 6

Signal Gain and DC Offset
Element-Wise Processing: Scalar Operations with Arrays

6.1 Introduction: Digital Signal Processing

Digital signal processing (DSP) is a set of techniques that can be used to change the information in a signal. DSP involves any method of processing or analyzing a signal to create a new signal or to recover information about the original signal. Applications of DSP are widespread and have many uses in audio for sampled signals. This chapter introduces techniques related to processing audio signals with operations using scalar numbers.

6.1.1 Element-Wise Processing

To begin, a specific category of processing is presented called **element-wise processing**. In this type of processing, the output at a given sample (or element) depends only on the input at the same sample. As an example, for signals $x = \{x_1, x_2, ..., x_N\}$ and $y = \{y_1, y_2, ..., y_N\}$, the command $y = x$ assigns each individual element of x to the corresponding element of y, such that $y_1 = x_1$, $y_2 = x_2$, ..., $y_N = x_N$.

Besides simply assigning the input directly to the output, there are many systems that apply various linear operations to the input element before assigning it to the output element. These operations are the focus of the current chapter. Then, this same approach is used for signal synthesis in Chapter 7, digital summing in Chapter 8, stereo panning in Chapter 9, and distortion effects in Chapter 10.

Chapter 11 introduces other systems for processing audio not based only on element-wise processing, categorized as *systems with memory*.

6.1.2 Element-Wise Referencing

For element-wise processing, it is desired to reference (or index) the elements of an input array and assign each to the corresponding elements of an output array. One explicit method to accomplish this processing is to use a loop (Section 5.3.2) to step through the individual elements of an array. Each iteration of the loop references an element of the input array and assigns it to the same reference of the output array. After every element has been referenced, the task is complete.

Examples: Create and run the following m-file in MATLAB.

```
% ELEMENTLOOP
% This script demonstrates a method to reference
% and assign the elements of an input array to an
% output array
%
% A loop is used to step through each element of
% the arrays

clear; clc;
% Example - Sine wave test signal
filename = 'sw20Hz.wav';
[x,Fs] = audioread(filename); % input signal

N = length(x);
y = zeros(N,1);
% Initialize an empty output array
% Loop through arrays to perform element-wise referencing
% n - variable for sample number
for n = 1:N
    % Note: a new element from "x" is assigned to "y"
    % each time through the loop
    y(n,1) = x(n,1)

end
```

Example 6.1: Script: `elementLoop.m`

6.1.3 Block Diagrams

One way to visualize the processing of a signal is by using a **block diagram**. A block diagram represents how an *input* signal becomes an *output* signal after being routed through various processing blocks. Block diagrams use arrows to represent how a signal is routed from start to finish. A basic block diagram, shown in Figure 6.1, displays the input signal, x, routed to the

Figure 6.1: Basic block diagram ·

output signal, y. As demonstrated in Example 6.1, one approach to implementing this
processing uses a loop to explicitly assign each sample value of the input signal to each
sample value of the output signal. An alternative approach is possible in MATLAB using a
single command, y = x, to assign the entire input array to the entire output array.

Additonal block diagrams are used throughout this chapter to visualize systems that process
signals. Besides arrows for routing signals, symbols and blocks can be incorporated into the
diagram for mathematical operations and other processing functions.

6.2 *Scalar Operations with Arrays*

Arrays can be added, subtracted, multiplied, divided, and raised to the power of a scalar
(single numbers). In this case, the scalar is applied to each element of the array individually.
This operation is accomplished using conventional mathematical symbols.

Examples: Execute the following commands in MATLAB.

```
>> x = [ 1 2 3 ];
>> addSc = x + 1
>> subSc = x - 1
>> multSc = x * 2
>> divSc = x / 2
>> powErr = x ^ 2
>> powSc = x .^ 2
>> freqSpacing = 2 * logspace(1,4,10)
```

Example 6.2: Array operations with scalars

There are many mathematical operations in MATLAB that can be used directly with arrays.
One of the distinguishing differences between MATLAB and other programming languages is
the MATLAB library of functions and operations for working with arrays. Only a subset
of these operations are presented based on their relevance to the introductory scope of this
text. More information can be found in the MATLAB help documentation for *Matrix
Operations*.

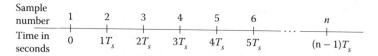

Figure 6.2: Time units of a signal

6.2.1 Scalar Multiplication: Converting Time Units

One use of scalar multiplication is to convert an array with time units of samples to an array with time units of seconds. The sampling period, T_s, can be used as the conversion factor:

$$samples \cdot \frac{T_s \; seconds}{sample} = seconds$$

As a convention, a signal begins at a time of 0 seconds. Due to MATLAB array indexing starting at element 1, the relationship between sample, n, and time, t, is $t = (n-1) \cdot T_s$. A timeline demonstrating the relationship between sample number and time in seconds is shown in Figure 6.2.

Examples: Execute the following commands in MATLAB.

```matlab
% This script demonstrates two approaches of converting a
% time in units of samples to a time in units of seconds.
% First, it is accomplished one-by-one inside a loop. Second,
% it is accomplished all at once using array multiplication.

% Import Sound File
filename = 'sw20Hz.wav';
[x,Fs] = audioread(filename);
Ts = 1/Fs;     % Determine sampling period (sec/sample)
N = length(x); % Total number of samples

% Method 1 - Inside a Loop
for n = 1:N
    % Convert sample number "n" to units of seconds "t"
    t = (n-1) * Ts     % sec = sample * (sec/sample)
    % Note: MATLAB indexing starts at n = 1
    % As a convention, time in seconds starts at t = 0
end


% Method 2 - Array Multiplication
```

```
% In this case, an array of sample numbers is created [0:N-1].
% Then, it is multiplied by the sampling period to create a
% time vector with units of seconds.
% sec = sample * (sec/sample)
t = [0:N-1]*Ts;
% Rotate "t" to a column vector
t = t(:);
```

Example 6.3: Convert samples to units of seconds

6.3 Scalar Multiplication: Signal Gain

The amplitude of a signal can be changed by multiplying each element by a scalar value, g. An increase to the amplitude of a signal can be accomplished by multiplying the signal by a value greater than one. A decrease to the amplitude of a signal can be accomplished by multiplying the signal by a value less than one.

A closely related operation is element-wise division. For a signal, x, and divisor, g, element-wise division is calculated using the expression x / g. This operation can be performed equivalently by multiplying by the inverse of the divisor: x * (1/g).

6.3.1 Signal Gain Block Diagram

A block diagram is shown in Figure 6.3 to represent the process of multiplying an input signal by a gain value, g.

The process to scale a signal's amplitude is the following. The output signal, $y = \{y_1, y_2, ...,$ $y_N\}$, is a function of the input signal, $x = \{x_1, x_2, ..., x_N\}$, and the gain value, g. Each individual element of y is calculated from the corresponding element of x such that $y_1 = g * x_1$, $y_2 = g * x_2, ... , y_N = g * x_N$. In general, the process can be written $y[n] = g * x[n]$. In MATLAB, the process can be executed by the command y = g * x.

In Example 6.4, a signal's amplitude is scaled by using two different approaches. First, each individual sample of the input signal is referenced, scaled, and assigned to the corresponding

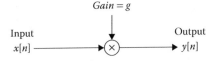

Figure 6.3: Block diagram with multiplication for gain scaling

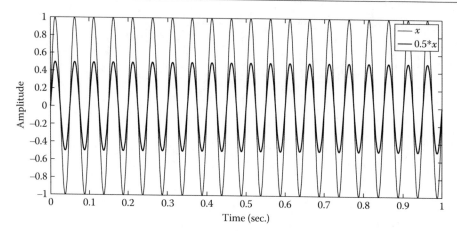

Figure 6.4: Scaling the amplitude of a signal

sample of the output signal. Second, the signal's amplitude is scaled by using the array operation with the scalar. In Figure 6.4, a plot of the result is shown.

Examples: Create and run the following m-file in MATLAB.

```
% SCALEAMP
% This script demonstrates two methods for
% scaling the amplitude of a signal. The first
% method uses a loop to perform element-wise
% indexing of the input signal. The second method
% uses MATLAB's scalar operation with the array.

clear; clc;

% Import the input signal
filename = 'sw20Hz.wav';
[x,Fs] = audioread(filename); % input signal

Ts = 1/Fs;
% Time vector for plotting
t = [0:length(x)-1]*Ts; t = t(:);

% Example 1 - Loop
g1 = 0.5; % Gain Scalar
```

```
N = length(x);
y1 = zeros(N,1);
% n - variable for sample number
for n = 1:N
    % Multiply each element of "x" by "g1"
    y1(n,1) = g1 * x(n,1);

end

figure(1); % Plot Example 1
plot(t,x,'--',t,y1); legend('x','0.5*x');

% Example 2 - Array Operation
g2 = 0.25;

% In this approach, it is not necessary to use
% a loop to index the individual elements of "x".
% By default, this operation performs element-wise processing
y2 = g2 * x;

figure(2); % Plot Example 2
plot(t,x,t,y2);
```

Example 6.4: Script: `scaleAmp.m`

6.3.2 Polarity Inversion

A noteworthy case of scalar multiplication is when a value of -1 is multiplied by each sample of a signal. This operation results in the positive signal values becoming negative, and the negative signal values becoming positive. In audio, this operation is referred to as **polarity inversion**. Many audio consoles, as well as software processors, include a function to invert a signal's polarity by pressing a button labeled $\boxed{\emptyset}$. An example of the block diagram for polarity inversion is shown in Figure 6.5.

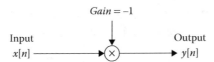

Figure 6.5: Block diagram of polarity inversion

Examples: Create and run the following m-file in MATLAB.

```
% This script demonstrates the process of inverting the polarity
% of a signal. It is performed by multiplying the signal by -1.


% 20 Hz signal for visualization
filename = 'sw20Hz.wav';
[sw,Fs] = audioread(filename);
Ts = 1/Fs;        % Sampling period
N = length(x);    % Total number of samples in signal
t = [0:N-1]*Ts; t=t(:);  % Corresponding time vector in seconds
% Polarity Inversion
y = -1 * x;
plot(t,x,t,y); % Plot the original signal and the processed signal.
% The processed signal should be a "mirror image" version of the
% original reflected across the horizontal axis.
```

Example 6.5: Inverting the polarity of a signal

Audio engineers have casually referred to this process as "flipping the phase" of a signal. However, it is more accurately described as a change in polarity (i.e., convention defining positive and negative amplitude) and not a change in phase (i.e., relative point in a cycle's period).

6.3.3 Decibel Scale

The **decibel (dB) scale** is used extensively in audio for processing the amplitude of a signal. It is a logarithmic transformation of the linear scale. Both scales can be used to describe the amplitude of a signal. In some ways, it is similar to how temperature can be described using the Celsius scale or the Fahrenheit scale. When programming a computer for audio, it is necessary to be able to convert from one scale to the other.

Background

In many audio applications, the amplitude of a signal, A_1, is compared to another amplitude, A_2. As one example, signal amplitude, A_{sig}, could be compared to an absolute reference, A_{ref}. As another example, the output amplitude, A_{out}, of a signal from an effect could be compared to the input amplitude, A_{in}.

The difference in amplitude, $A_1 - A_2$, is one way to perform the comparison. Using the definition of signal power, $P = A^2$, a related comparison is $P_1 - P_2$.

Alexander Graham Bell proposed using a logarithmic comparison, $log_{10}(P_1) - log_{10}(P_2)$, with scientific units: *Bel*. Transforming this expression for signal amplitude leads to the following mathematical relationships:

$$log_{10}(P_1) - log_{10}(P_2) \, \text{Bel} = log_{10}\left(\frac{P_1}{P_2}\right) \text{Bel}$$

$$log_{10}\left(\frac{P_1}{P_2}\right) \text{Bel} \cdot \frac{10 \text{ decibel}}{1 \text{ Bel}} = 10 \cdot log_{10}\left(\frac{P_1}{P_2}\right) \text{dB}$$

$$= 10 \cdot log_{10}\left(\frac{A_1^{\,2}}{A_2^{\,2}}\right) \text{dB}$$

$$= 20 \cdot log_{10}\left(\frac{A_1}{A_2}\right) \text{dB}$$

Decibel Scale Processing

When processing the amplitude of audio signals, it is common to change the amplitude relative to a decibel scale. Since the amplitude of a signal is changed by multiplying by a scaling factor on a linear scale, it is necessary to convert the desired change on a decibel scale to the equivalent scaling factor on a linear scale.

The following equation can be used to convert a desired change in amplitude on a decibel scale, A_{dB}, to a scaling factor on a linear scale, A_{lin}:

$$A_{lin} = 10^{\left(\frac{A_{dB}}{20}\right)} \tag{6.1}$$

To determine the resulting change in amplitude on a decibel scale, A_{dB}, by a scaling factor on a linear scale, A_{lin}, use the following:

$$A_{dB} = 20 \cdot log_{10}\left(\frac{A_{lin}}{1}\right) \tag{6.2}$$

Examples: Save the following function as an m-file in MATLAB.

```
% DBAMPCHANGE This function changes the amplitude of a signal
% This function changes the amplitude of an input signal based on a
% desired change relative to a decibel scale

function [ out ] = dbAmpChange( in,dBChange )

scale = 10^(dBChange/20); % Convert from decibel to linear
```

```
out = scale*in;          % Apply linear amplitude to signal



end
```

Example 6.6: Function: `dbAmpChange.m`

Examples: Create and run the following m-file in MATLAB.

```
% DBAMPEXAMPLE
% This script provides an example for changing the amplitude
% of a signal on a decibel (dB) scale
%
% See also DBAMPCHANGE

% Example - Sine wave test signal
[sw1,Fs] = audioread('sw20Hz.wav');
sw2 = dbAmpChange(sw1,6);
sw3 = dbAmpChange(sw1,-6);

% Plot the result
plot(t,sw1,'.',t,sw2,t,sw3,'--');
xlabel('Time (sec.)');
ylabel('Amplitude');
title('Changing the Amplitude of a Signal on a dB Scale');
legend('SW1','+6 dB','-6 dB');
```

Example 6.7: Script: `dbAmpExample.m`

6.4 Visualizing the Amplitude Change

One method to visualize a change in signal amplitude is to view the waveform (Section 3.4). A comparison can be made between the original, unprocessed signal and the new, processed signal. In this method, the amplitude is plotted over time. Therefore, the gain change is easier to view for signals with a relatively consistent amplitude.

6.4.1 Input versus Output Characteristic Curve

Another method to visualize the processing of an element-wise system is to plot the amplitude of the input signal (on the horizontal axis) versus the amplitude of the output signal (on the vertical axis). This visualization is called the **characteristic curve**, also known in audio as the transfer curve, the transfer relationship, the waveshaper curve, or the I/O relationship. In Example 6.8, a method to analyze the characteristic curve for a scalar gain change is demonstrated. In addition to the examples shown in this chapter for linear processing, the characteristic curve is used extensively in Chapter 10 for nonlinear processing.

Examples: Create and run the following m-file in MATLAB.

```matlab
% CHARACTERISTICCURVE
% This script provides two examples for analyzing the
% characteristic curve of an audio effect which uses
% element-wise processing
%
% This first example creates an input array with values
% from -1 to 1. The second example uses a sine wave signal

clear;clc;close all;

% Example 1: Array [-1 to 1] to span entire full-scale range
in = [-1:.1:1].';

% Example 2: Sine Wave Test Signal. This example shows the
% characteristic curve can be created using any signal
% which spans the full-scale range, even if it is periodic.
% Uncomment this code to switch examples.
% [in,Fs] = audioread('sw20Hz.wav');

% Assign in to out1, for comparison purposes
out1 = in;    % No amplitude change

N = length(in);
out2 = zeros(N,1);
% Loop through arrays to perform element-wise multiplication
for n = 1:N
    out2(n,1) = 2*in(n,1);
end

% Element-wise multiplication can also be
% accomplished by multiplying an array directly by a scalar
out3 = 3*in;

% Plot the Characteristic Curve (In vs. Out)
```

```
plot(in,out1,in,out2,in,out3);
xlabel('Input Amplitude');
ylabel('Output Amplitude');
legend('out1 = in','out2 = 2*in','out3 = 3*in');
% Draw axes through origin
ax = gca;
ax.XAxisLocation = 'origin';
ax.YAxisLocation = 'origin';
```

Example 6.8: Script: `characteristicCurve.m`

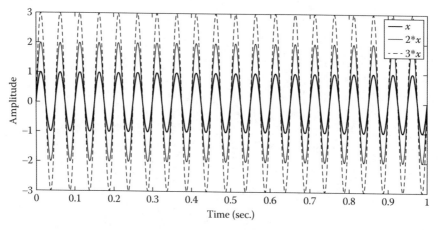

Figure 6.6: Waveform of signal gain change

In Figure 6.6, the waveform of the processed signals are shown. In Figure 6.7, the processing is visualized using the characteristic curve. For values of the input signal shown along the horizontal axis, the corresponding value of the output signal is shown by the plotted line. By performing scalar multiplication the line changes, which indicates the output signal would have different values.

6.5 Scalar Addition: DC Offset

Element-wise scalar addition can be used to equally shift all elements of a signal relative to an amplitude value of 0. This is commonly described as a DC offset. This is the equivalent of adding a signal that does not oscillate (0 Hz), named for analog direct current (DC) signals.

6.5.1 DC Offset Block Diagram

A block diagram is shown in Figure 6.8 to represent the process of adding a DC offset, μ, to an input signal.

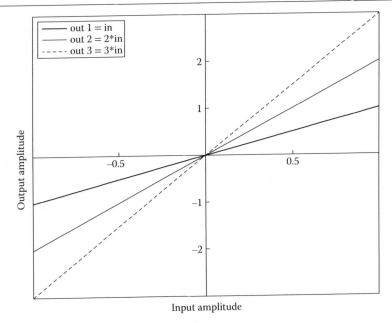

Figure 6.7: Characteristic curve for a gain change

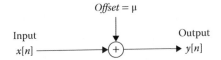

Figure 6.8: Block diagram with addition for DC offset

The process to add a DC offset to a signal is the following. The output signal, $y = \{y_1, y_2, ..., y_N\}$, is a function of the input signal, $x = \{x_1, x_2, ..., x_N\}$, and DC offset, μ. Each individual element of y is calculated from the corresponding element of x such that $y_1 = x_1 + \mu$, $y_2 = x_2 + \mu$, ... , $y_N = x_N + \mu$. In general, the process can be written $y[n] = x[n] + \mu$. In MATLAB, the process can be executed by the command $y = x + \mu$. In Example 6.9, a method to analyze the characteristic curve for the DC offset process is demonstrated. In Figures 6.9 and 6.10, the resulting waveforms and characteristic curves are shown, respectively.

Examples: Create and run the following m-file in MATLAB.

```
% DCOFFSET
% This script demonstrates a method to perform
% element-wise scalar addition, the equivalent
% of a DC offset

clear;clc;close all;
```

```matlab
% Example - Sine Wave Signal
filename = 'sw20Hz.wav';
[in,Fs] = audioread(filename); % Import sound file
Ts = 1/Fs;
% Assign in to out1, for comparison purposes
out1 = in;

N = length(in);
out2 = zeros(N,1); % Initialize output array
% Loop through arrays to perform element-wise scalar addition
for n = 1:N
    out2(n,1) = in(n,1)+1;
end

% Element-wise addition can also be
% accomplished by adding a scalar directly to an array
out3 = in-0.5;

figure; % Create a new figure window
% Plot the output amplitude vs. time
t = [0:N-1]*Ts; t = t(:);
plot(t,out1,t,out2,t,out3);
xlabel('Time (sec.)');
ylabel('Amplitude');
legend('out1 = in','out2 = in+1','out3 = in-0.5');

figure;
% Plot the input vs. output
plot(in,out1,in,out2,in,out3);
xlabel('Input Amplitude');
ylabel('Output Amplitude');
legend('out1 = in','out2 = in+1','out3 = in-0.5');
% Draw axes through origin
ax = gca;
ax.XAxisLocation = 'origin';
ax.YAxisLocation = 'origin';
```

Example 6.9: Script: dcOffset.m

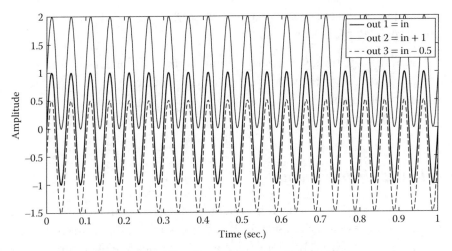

Figure 6.9: Waveform of signals with DC offset

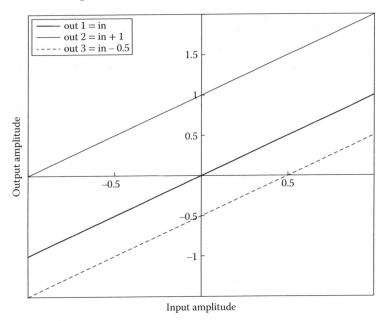

Figure 6.10: Characteristic curve with DC offset

6.6 Combined Signal Gain and DC Offset

In the combined case, the output signal, $y = \{y_1, y_2, ..., y_N\}$, is a function of the input signal, $x = \{x_1, x_2, ..., x_N\}$, gain value, g, and DC offset, μ. Each individual element of y is calculated from the corresponding element of x such that $y_1 = g * x_1 + \mu$, $y_2 = g * x_2 + \mu$, ... ,

$y_N = g * x_N + \mu$. In general, the process can be written $y[n] = g * x[n] + \mu$. In MATLAB, the process can be executed by the command $y = g * x + \mu$.

This result is analogous to the generalized algebraic form of linear equations, $y = mx + b$, where the slope is determined by m and the y-intercept is determined by b. As shown by the characteristic curve in Figure 6.11, the gain value g changes the slope of the line and the DC offset μ sets the y-intercept.

6.7 *Amplitude Measurements*

There are several different ways to measure and describe the overall amplitude of a signal. Each measurement represents an amplitude characteristic for the entire signal, as opposed to one of the sample values, $x[n]$. Therefore, these amplitude measurements are not functions over time (or sample number).

A signal's peak amplitude, A_{pk}, is a measure of its maximum instantaneous amplitude. A variation of this measurement is the peak-to-peak amplitude, A_{pp}, which describes the difference between the maximum and minimum instantaneous amplitude. A signal's root-mean-square (RMS) amplitude, A_{rms}, is a measure of its average amplitude across all samples. A visualization of each measurement is shown in Figure 6.12.

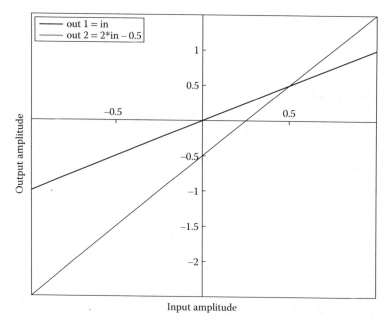

Figure 6.11: Characteristic curve with gain and DC offset

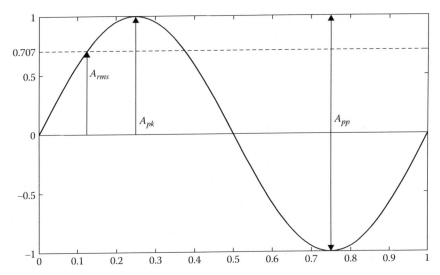

Figure 6.12: Measurements of peak, peak-to-peak, and RMS amplitude

6.7.1 Peak Amplitude

To measure a signal's **peak amplitude**, it is necessary to determine the sample value with the maximum amplitude. Because audio signals can have positive and negative values, a signal's peak magnitude is typically used for peak amplitude. Magnitude is a measure of the distance or deviation from zero, and is strictly positive. Therefore, the peak magnitude is the sample with the greatest difference from zero, regardless of whether the sample has a positive or negative value.

The absolute value function, abs(), can be used to take the magnitude of a signal. Then, the function max() can be used to determine which sample has the greatest magnitude. For a signal, $x = \{x_1, x_2, ..., x_N\}$, the peak amplitude, A_{pk}, is:

$$A_{pk} = \texttt{max(abs(x))}$$

The peak amplitude for a digital signal can be compared relative to the maximum digital amplitude of 1, called the full-scale (FS) amplitude. When needed, this value can be converted to units of decibel full scale (dBFS) using: $20 \cdot log_{10}(\frac{A_{pk}}{1})$.

6.7.2 Peak-to-Peak Amplitude

To measure a signal's **peak-to-peak amplitude**, the difference between the maximum and minimum amplitude values is calculated. In this case, it is not worthwhile to consider the magnitude of a signal's amplitude. For typical audio signals with positive and negative amplitude values, this measurement finds the difference between the sample with the highest positive value and the sample with the highest negative number. For a signal, $x = \{x_1, x_2, ..., x_N\}$, the peak-to-peak amplitude, A_{pp}, is:

$$A_{pp} = \max(x) - \min(x)$$

6.7.3 Root-Mean-Square Amplitude

If the peak magnitude is a measure of instantaneous amplitude, the **root-mean-square (RMS) amplitude** is a measurement of average signal magnitude. Peak amplitude measures the single sample in the signal with the greatest magnitude, regardless of the amplitude of any other samples. For the RMS measurement, the amplitude of every sample in the signal contributes to the result. For this reason, the RMS amplitude can be used as a primative approximation of how a listener perceives the strength of a signal.

RMS Definition

Although the RMS amplitude represents the average signal strength, the arithmetic mean of the sample values cannot be used. For a typical audio signal with positive and negative amplitude values, the arithmetic mean would be zero (or very close to zero). Essentially, the positive sample values would cancel with the negative sample values. This would be true of signals with a small peak-to-peak amplitude and signals with a large peak-to-peak amplitude. Therefore, the arithmetic mean is not a valid calculation of average signal strength. Instead, the RMS amplitude is calculated using the square *root* of the arithmetic *mean* of the *squares* of the individual sample values. For a signal, $x = \{x_1, x_2, ..., x_N\}$, the RMS amplitude, A_{rms}, is:

$$A_{rms} = \sqrt{\frac{1}{N}(x_1{}^2 + x_2{}^2 + ... + x_N{}^2)}$$

Examples: Create and run the following m-file in MATLAB.

```
% This script demonstrates the process of calculating the RMS
% amplitude of a signal. First, the samples in the signal are
% squared. Second, the mean (arithmetic average of the squared
% signal is found. Third, the square root of the mean of the
% squared signal is determined.

% Sine wave signal for testing
[x] = audioread('sw20Hz.wav');
N = length(x); % Total number of samples


% Square the individual samples of the signal
```

```
sigSquared = x.^2;     % Element-wise power operation
                       % Result is still an array


% Find the mean. Note: result is now a scalar
sigMeanSquared = (1/N) * sum(sigSquared);


% Take the square root
sigRootMeanSquared = sqrt(sigMeanSquared)
% The result printed to the command window is the
% RMS amplitude for a sine wave => 0.707
```

Example 6.10: Calculating the RMS amplitude of a signal

Summation Notation

The following notation can be used to condense the written form for the summation of multiple values:

$$\text{Let: } \sum_{n=1}^{N} n = 1 + 2 + 3 + \ldots + N$$

$$\sum_{n=1}^{N} x_n = x_1 + x_2 + x_3 + \ldots + x_N$$

$$\sum_{n=1}^{N} (x_n)^2 = x_1^2 + x_2^2 + x_3^2 + \ldots + x_N^2$$

Therefore, the RMS amplitude is written using the summation notation as:

$$A_{rms} = \sqrt{\frac{1}{N} \sum_{n=1}^{N} (x_n)^2} \tag{6.3}$$

Full-scale RMS Amplitude

For musical signals, the full-scale (FS) reference value for RMS amplitude is based on a sine-wave signal. This is a standard convention adopted by the *Audio Engineering Society* (AES7-2015). Therefore, the RMS amplitude for a signal can be converted to units of decibel full scale (dBFS) using $20 \cdot log_{10}\left(\frac{A_{rms}}{.707}\right)$.

6.7.4 Dynamic Range Crest Factor

One way to analyze the relative loudness of a signal is to compare the maximum signal level (peak amplitude) and the average signal level (RMS amplitude). A signal's **crest factor** (CF) is defined as the ratio of peak amplitude to RMS amplitude, also known as the **peak-to-average ratio** (PAR) (AES2-2012).

$$CF = \frac{|A_{pk}|}{A_{rms}} \qquad (6.4)$$

During the process of mastering a musical recording, dynamic range reduction with a compressor (Section 18.1.2) or limiter is applied to the recorded signal. This audio effect causes the crest factor to be decreased because A_{rms} is increased without changing A_{pk}. Therefore, the crest factor is an analysis used to describe the relative amount of dynamic range (DR) in a signal.

A common transformation of the calculation is to consider the crest factor on the decibel scale:

$$CF_{dB} = 20 \cdot log_{10}(|A_{pk}|) - 20 \cdot log_{10}\left(\frac{A_{rms}}{0.707}\right) \qquad (6.5)$$

The result is used as the value of crest factor DR for a mastered musical recording. Typical values range from DR5 (less dynamic range) to DR15 (more dynamic range).

6.8 Amplitude Normalization

Amplitude normalization is the process of scaling the amplitude of a signal to have a specified value. There are two types of amplitude normalization, peak and RMS, chosen to scale a signal's amplitude to the respective measure.

6.8.1 Peak Normalization

Peak normalization is the process by which the amplitude of a signal is scaled such that the signal peak reaches a specified value. For a signal, $x[n] = \{x_1, x_2, ..., x_N\}$, let $\tilde{x}[n]$ be the processed, peak-normalized signal. Then, a signal can be peak normalized to full-scale amplitude using the following expression:

$$\tilde{x}[n] = \frac{x[n]}{A_{pk}}$$

Furthermore, the peak amplitude of a signal can be normalized to any specified linear value, ρ, using the expression:

$$\tilde{x}[n] = \rho \cdot \frac{x[n]}{A_{pk}}$$

If a desired peak amplitude is known on the decibel scale, ρ_{dB}, it can be converted to a linear value with $\rho = 10^{\frac{\rho_{dB}}{20}}$.

6.8.2 *Root-Mean-Square Normalization*

Root-mean-square (RMS) normalization is the process by which the amplitude of a signal is scaled such that the signal's RMS amplitude is a specified value. A signal, $x = \{x_1, x_2, ..., x_N\}$, can be normalized to a specified RMS amplitude, ρ, by multiplying the amplitude of each sample by a scaling factor, α. The scaling factor can be determined for the specified RMS amplitude using the following relationship:

$$\rho = \sqrt{\frac{1}{N}[(\alpha x_1)^2 + (\alpha x_2)^2 + ... + (\alpha x_N)^2]}$$

$$\rho^2 = \frac{1}{N}[(\alpha x_1)^2 + (\alpha x_2)^2 + ... + (\alpha x_N)^2]$$

$$N \cdot \rho^2 = \alpha^2(x_1{}^2 + x_2{}^2 + ... + x_N{}^2)$$

$$\frac{N \cdot \rho^2}{(x_1{}^2 + x_2{}^2 + ... + x_N{}^2)} = \alpha^2$$

$$\sqrt{\frac{N \cdot \rho^2}{(x_1{}^2 + x_2{}^2 + ... + x_N{}^2)}} = \alpha$$

Then, the RMS normalized signal, $\tilde{x}[n]$, can be calculated using $\tilde{x}[n] = \alpha \cdot x[n]$

Changing the RMS amplitude relative to another signal

There are circumstances when it is desired to change the amplitude of one signal relative to the RMS amplitude of another signal. As an example, an audio engineer might add background noise to a signal, such that the RMS amplitude of the noise is 12 dB less than the RMS amplitude of the signal. In this case, it would be necessary to measure the RMS amplitude for the signal, then scale the amplitude of the noise so its RMS amplitude is the desired value.

In general, assume there is a signal, $x[n] = \{x_1, x_2, ..., x_N\}$, with RMS amplitude x_{rms}. The amplitude of another signal, $y[n] = \{y_1, y_2, ..., y_M\}$, can be scaled by some amount, α, relative to x_{rms} to produce a processed signal, $\tilde{y}[n] = \{\alpha \cdot y_1, \alpha \cdot y_2, ..., \alpha \cdot y_M\}$. It is desired for the RMS amplitude of the processed signal, $\tilde{y}_{rms}$, to be related to x_{rms} based on the decibel value, ρ_{dB}, such that $10^{\left(\frac{\rho_{dB}}{20}\right)} \cdot x_{rms} = \tilde{y}_{rms}$.

The scaling factor, α, can be determined for signal $y[n]$ to meet the desired requirement by using the following relationships:

$$10^{\left(\frac{\rho_{dB}}{20}\right)} \cdot x_{rms} = \tilde{y}_{rms}$$

$$\text{Let: } \rho = 10^{\left(\frac{\rho_{dB}}{20}\right)}$$

$$\rho\sqrt{\frac{1}{N}(x_1^2 + x_2^2 + ... + x_N^2)} = \sqrt{\frac{1}{M}((\alpha y_1)^2 + (\alpha y_2)^2 + ... + (\alpha y_M)^2)}$$

$$\rho^2\frac{1}{N}(x_1^2 + x_2^2 + ... + x_N^2) = \frac{1}{M}((\alpha y_1)^2 + (\alpha y_2)^2 + ... + (\alpha y_M)^2)$$

$$\frac{M\rho^2}{N}(x_1^2 + x_2^2 + ... + x_N^2) = \alpha^2(y_1^2 + y_2^2 + ... + y_M^2)$$

$$\frac{M\rho^2}{N}\frac{(x_1^2 + x_2^2 + ... + x_N^2)}{(y_1^2 + y_2^2 + ... + y_M^2)} = \alpha^2$$

$$\sqrt{\frac{M\rho^2}{N}\frac{(x_1^2 + x_2^2 + ... + x_N^2)}{(y_1^2 + y_2^2 + ... + y_M^2)}} = \alpha$$

Bibliography

AES2-2012 standard, 3.14 Crest factor, *AES standard for acoustics – Methods of measuring and specifying the performance of loudspeakers for professional applications – Drive units*, p. 6 in the February 11, 2013 printing. New York: Audio Engineering Society.

AES7-2015 standard, 3.12.3 Decibels full scale, *AES standard method for digital audio engineering – Measurement of digital audio equipment*, p. 5 in the July 17, 2015 printing. New York: Audio Engineering Society.

Chapter 7

Introduction to Signal Synthesis
Element-Wise Processing: Built-In Functions with Arrays

7.1 Introduction: Signal Synthesis

Audio signals can be synthesized by using mathematical functions. There are many basic signals used in audio as "test" sound waves or with synthesizers. This chapter explores how an array of time samples can be processed by a built-in function to create various signals. These signals include a sine wave pure tone, square wave, sawtooth wave, triangle wave, impulse train, and white noise. Therefore, both periodic and aperiodic signals are discussed.

7.2 Visualizing the Frequency Spectrum of an Audio Signal

There are many ways audio engineers can visualize the differences in various signals. In Section 3.4, the waveform plot of a signal's amplitude versus time is discussed. Another way to visualize a signal is called the **frequency spectrum**. It is a plot of a signal's amplitude versus frequency, typically between 20–20,000 Hz.

An audio signal contains one, or more, frequencies at a given time. It can be complicated to determine which frequencies are contained in a signal based on the waveform. The frequency spectrum provides a way to see the amplitude of separate, individual frequencies.

A MATLAB function is included with the book's *Additional Content* to plot a signal's waveform and frequency spectrum. The syntax to use the function is `plottf(x,Fs)`, where x is the signal and Fs is its sampling rate. This function produces a figure window with the waveform plotted on the top and the frequency spectrum plotted on the botton.

Examples: Execute the following commands in MATLAB.

```
>> [x,Fs] = audioread('sw440Hz.wav');
>> plottf(x,Fs);
```

Example 7.1: Function: `plottf`

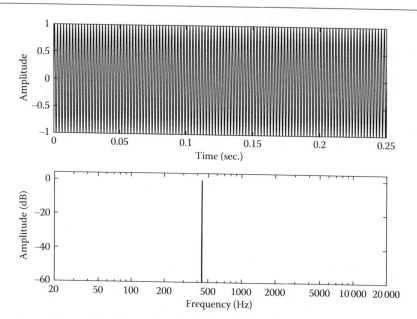

Figure 7.1: Waveform and frequency spectrum of a sine wave

In Figure 7.1, the waveform and frequency spectrum of a 440 Hz sine wave is plotted. The frequency spectrum shows this sine wave has a peak amplitude of unity gain (0 dBFS) at 440 Hz, and does not contain any other frequencies.

7.3 Periodic Signals

There are many audio signals that repeat a pattern of motion. Each repetition is called a cycle, and the time length of a cycle is called the period. Therefore, signals with this behavior are **periodic signals**.

Each position within a cycle is defined as an angle of rotation called the **phase angle**. Units of rotation are defined as degrees $[0° - 360°]$ or radians $[0 - 2\pi]$. The unit of radians is defined as the arc length for the unit circle using an angle of rotation. The relationship between phase angle in degrees and radians on the unit circle along with Cartesian coordinates (x, y values) is shown in Figure 7.2 (Aryal, 2010).

7.3.1 Sine Wave

One basic periodic signal is the **sine wave**. It represents the motion in a single dimension (consider the motion of a speaker cone) for a signal with a phase angle that rotates at a constant rate. Therefore, a sine wave is a signal with a single frequency.

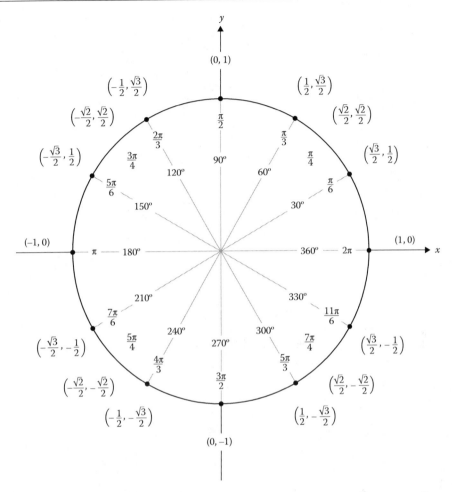

Figure 7.2: Unit circle relationship between degrees, radians, and Cartesian coordinates

Trigonometric Sine Function

A sine wave signal is based on the trigonometric sine function. The sine function is defined over the period of a single cycle. On the unit circle, the sine of a phase angle θ is defined as the ratio of the length of the opposite and hypotenuse in a right triangle, shown in Figure 7.3.

For a unit circle with radius (hypotenuse) equal to one, $sin(\theta)$ is equal to the y value (motion along the vertical dimension) in Cartesian coordinates where the hypotenuse intersects the circle, shown in Figure 7.4.

Synthesizing a Sine Wave Signal

A sine wave signal can be synthesized using the built-in function `sin(x)`. Rather than calculating the sine function for a single value, x, the function is used to perform element-wise

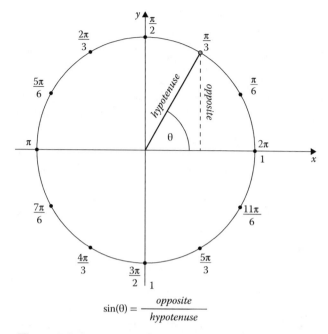

$$sin(\theta) = \frac{opposite}{hypotenuse}$$

Figure 7.3: Unit circle relationship with the sine function

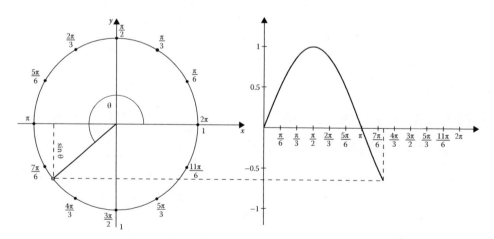

Figure 7.4: Unit circle relationship with the sine waveform

processing on a time vector, $t = \{t_1, t_2, \ldots, t_n\}$, with units of seconds. In this case, $\mathtt{sin(t)}$ = $\{sin(t_1), sin(t_2), \ldots, sin(t_n)\}$. When synthesizing a signal over units of time (seconds) the relationships in Table 7.1 are true.

Table 7.1: Transforming the sine function

Function Input	Output Characteristics
$sin(t)$	$\frac{1\ cycle}{2\pi\ seconds}$
$sin(2\pi \cdot t)$	$\frac{1\ cycle}{1\ second}$
$sin(2\pi \cdot f \cdot t)$	$\frac{f\ cycles}{1\ second}$
$sin(2\pi \cdot f \cdot t + \phi)$	Phase offset, $\phi \in [0, 2\pi]$
$A \cdot sin(2\pi \cdot f \cdot t + \phi)$	Amplitude, A

Therefore, the sine wave signal can be written $x[t] = A \cdot sin(2 \cdot \pi \cdot f \cdot t + \phi)$, where f is a frequency scalar, t is a time vector of samples, and ϕ is a phase offset from $[0, 2\pi]$. In Example 7.2, two different methods to synthesize a sine wave are demonstrated. The result is plotted in Figure 7.5.

Examples: Create and run the following m-file in MATLAB.

```matlab
% SINESYNTHESIS
% This script demonstrates two methods to synthesize
% sine waves
%
% Method 1: A loop is used to step through each
% element of the arrays
%
% Method 2: Array processing is used to perform
% element-wise referencing by the "sin" function
% internally
%
% See also SINEANGLE, SINESPECTRUM

clear; clc;
% Example - Sine Wave Signal
% Declare initial parameters
f = 2; % frequency in Hz
phi = 0; % phase offset
Fs = 100; % sampling rate
Ts = 1/Fs; % sampling period
lenSec = 1; % 1 second long signal
N = Fs*lenSec;  % convert to time samples
out1 = zeros(N,1);
% Method 1: Loop to perform element-wise referencing
```

```
for n = 1:N
    % Note: a new time "t" is used
    % each time through the loop
    t = (n-1) * Ts;   % "t" is a scalar

    out1(n,1) = sin(2*pi*f*t+phi);

end

% Method 2: Create signal using array processing
% Phase shifted signal of identical frequency
t = [0:(N-1)] * Ts;  % sample times
t = t(:); % make a column vector
out2 = sin(2*pi*f*t+pi/2);   % 90 degree phase shift

% Plot the 2 signals
plot(t,out1,t,out2);
xlabel('Time (sec.)'); ylabel('Amplitude');
legend('out1','out2');
```

Example 7.2: Script: `sineSynthesis.m`

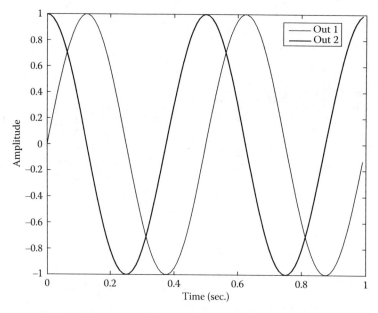

Figure 7.5: Waveform of sine waves with different phases

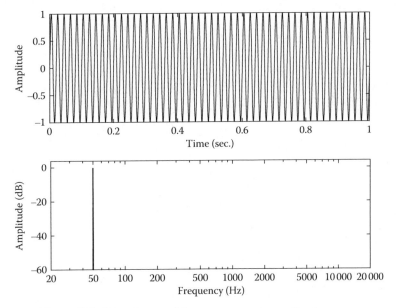

Figure 7.6: Waveform and frequency spectrum of a sine wave

In Figure 7.6, the waveform and frequency spectrum of a 50 Hz sine wave is plotted.

Phase Angle of Rotation Over Time

Another approach to synthesizing a sine wave signal is to reference the phase angle of rotation during the synthesis process. In this approach, the desired frequency of the signal is used to determine the necessary rate of rotation for a single sample. When the signal is synthesized, the phase angle is updated for each sample based on the rate of rotation. The following expression describes the relationship between frequency, f, and rate of rotation, $\Delta\Theta$:

$$\frac{f \text{ cycles}}{\text{sec}} \cdot \frac{T_s \text{ sec}}{\text{sample}} \cdot \frac{2\pi \text{ radians}}{\text{cycle}} = \frac{\Delta\Theta \text{ radians}}{1 \text{ sample}}$$

Examples: Create and run the following m-file in MATLAB.

```
% SINEANGLE
% This script demonstrates a method to synthesize
% sine waves by using an angle of rotation
%
% See also SINESYNTHESIS

clear;clc;
```

```
% Declare initial parameters
f = 2; % frequency in Hz
phi = 0; % phase offset
Fs = 1000; % sampling rate
Ts = 1/Fs; % sample period
t = [0:Ts:1].'; % sample times

% Calculate angle of rotation
angleChange = f*Ts*2*pi;
currentAngle = phi;

N = length(t);
out = zeros(N,1);
% Update the value of the currentAngle each iteration
% through the loop
for n = 1:N

    out(n,1) = sin(currentAngle);
    % Update phase angle for next loop
    currentAngle = currentAngle + angleChange;

    if currentAngle > 2*pi % Ensure angle is not > 2*pi
        currentAngle = currentAngle - 2*pi;
    end
end


% Plot the synthesized signal
plot(t,out);
xlabel('Time (sec.)'); ylabel('Amplitude');
legend('out');
```

Example 7.3: Script: `sineAngle.m`

7.3.2 Cosine Function

A related function to the sine function is the **cosine function**, $\cos(\theta)$. For an angular phase rotation, θ, the cosine function is the projection onto the horizontal axis of the unit circle. Therefore, the cosine function represents motion in a different dimension than the sine function. The relationship of the unit circle with the cosine funciton is shown in Figure 7.7.

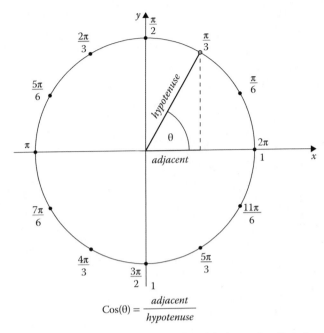

$$Cos(\theta) = \frac{adjacent}{hypotenuse}$$

Figure 7.7: Unit circle relationship with the cosine function

A cosine wave can be synthesized using the MATLAB function `cos(x)`. The function creates an output with some similarities to the sine function. Both the cosine function and the sine function can be used to create a signal with a single frequency. Comparing the two functions, the cosine function has a 90° $\left(\frac{\pi}{2}\right)$ phase difference with the sine function. Therefore, these functions are *orthogonal*, representing motion in perpendicular dimensions.

7.3.3 Square Wave

A **square wave** is a signal that oscillates between a single positive value and a single negative value. A square wave can be synthesized using the `square(x)` function in the Signal Processing Toolbox. In Example 7.4, a method to synthesize a square wave is demonstrated. The result is plotted in Figure 7.8.

Examples: Create and run the following m-file in MATLAB.

```
% SQUARESYNTHESIS
% This script demonstrates the "square" function
%
% See also SINESYNTHESIS, SAWTOOTHSYNTHESIS, TRIANGLESYNTHESIS

clear;clc;
```

```
% Example - Square Wave Signal
% 2 Hz square wave for visualization
f = 2;
phi = 0;
Fs = 1000;
Ts = (1/Fs);
t = [0:Ts:1].';
squareWave = square(2*pi*f.*t + phi);
plot(t,squareWave); axis([0 1 -1.1 1.1]);
xlabel('Time (sec.)');ylabel('Amplitude');
legend('2 Hz');

% 880 Hz square wave for audition
Fs = 44100;
Ts = (1/Fs);
t = 0:Ts:3; t = t(:);
f = 880;
sq = square(2*pi*f.*t);
sound(sq,Fs);
```

Example 7.4: Script: squareSynthesis.m

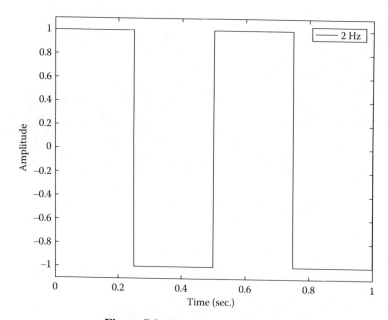

Figure 7.8: Creating a square wave

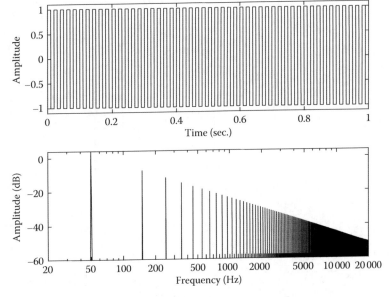

Figure 7.9: Waveform and frequency spectrum of a square wave

In Figure 7.9, the waveform and frequency spectrum of a 50 Hz square wave is plotted. As can be seen, a square wave is comprised of multiple frequencies. Specifically, there are harmonics at frequencies that are odd multiples of the fundamental frequency ($150 = 50 \cdot 3, 250 = 50 \cdot 5$, $350 = 50 \cdot 7$, etc.). The amplitude of the harmonics decreases with increasing frequency.

Square Wave Additive Synthesis

Taking into consideration its frequency spectrum, an approximation of a square wave can be created by combining multiple individual harmonics (i.e., sine functions). This method of forming a signal is called **additive synthesis**. The following equation can be used to describe a square wave as the summation of the odd harmonics of the fundamental frequency:

$$\text{Let: } x = 2 \cdot \pi \cdot f \cdot t$$

$$M = \left\lfloor \frac{F_s}{2 \cdot f} \right\rfloor$$

$$square(x) = \frac{4}{\pi} \left(sin(x) + \frac{1}{3} sin(3 \cdot x) + \frac{1}{5} sin(5 \cdot x) + \cdots \right)$$

$$square(x) = \frac{4}{\pi} \sum_{n=1,3,5,\ldots}^{M} \frac{1}{n} sin(n \cdot x)$$

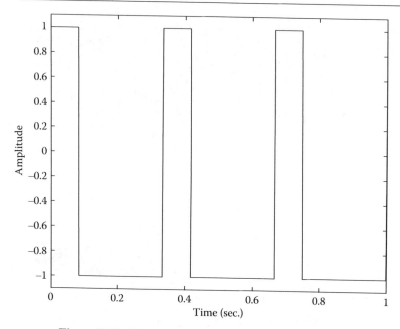

Figure 7.10: Changing the duty cycle of a square wave

The harmonic with the highest frequency is determined by the value M. It is calculated to be the number of odd harmonics less than the Nyquist frequency, $\frac{F_s}{2}$. This is done by rounding down ($\lfloor \cdot \rfloor$) to the nearest whole number of a harmonic less than the Nyquist frequency.

Duty Cycle

By default, the amplitude of a square wave is positive for half of the duration of the cycle and negative for the other half of the cycle. However, the length of positive and negative amplitude relative to the length of a cycle can be adjusted. The **duty cycle** is the percentage of one cycle in which the amplitude is positive. The following syntax can be used to set the duty cycle for a synthesized square wave: x = square(t,duty). The input variable, duty, is a number between [0,100]. In Example 7.5, a method to modify the duty cycle of a square wave is demonstrated. The result is plotted in Figure 7.10.

Examples: Create and run the following m-file in MATLAB.

```
% DUTYCYCLE
% This script synthesizes a square wave
% with an asymmetrical duty cycle
%
% See also SQUARESYNTHESIS
```

```
clear;clc;
% Example - Duty Cycle = 25
Fs = 1000;
Ts = (1/Fs);
t = 0:Ts:1;
t = t(:);
f = 3;
duty = 25;
sq = square(2*pi*f.*t, duty);
plot(t,sq); axis([0 1 -1.1 1.1]);
xlabel('Time (sec.)');ylabel('Amplitude');
```

Example 7.5: Script: `dutyCycle.m`

7.3.4 Sawtooth Wave

A **sawtooth wave** is a signal with amplitude that changes linearly between a minimum value and a maximum value. A sawtooth wave can be synthesized using the `sawtooth(x)` function in the Signal Processing Toolbox. In Example 7.6, a method to synthesize a sawtooth wave is demonstrated. The result is plotted in Figure 7.11.

Examples: Create and run the following m-file in MATLAB.

```
% SAWTOOTHSYNTHESIS
% This script demonstrates the "sawtooth" function
%
% See also SINESYNTHESIS, SQUARESYNTHESIS, TRIANGLESYNTHESIS

clear;clc;
% Example - Sawtooth Wave Signal
% 4 Hz signal for visualization
f = 4;
phi = 0;
Fs = 40;
Ts = (1/Fs);
t = [0:Ts:1].';
sawtoothWave = sawtooth(2*pi*f.*t + phi);
plot(t,sawtoothWave);

% 880 Hz signal for audition
Fs = 44100;
```

```
Ts = (1/Fs);
t = 0:Ts:3;
t = t(:);
f = 880;
st = sawtooth(2*pi*f.*t);
sound(st,Fs);
```

Example 7.6: Script: sawtoothSynthesis.m

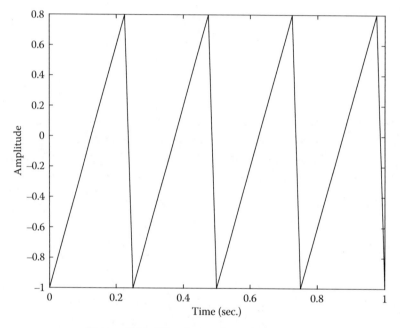

Figure 7.11: Waveform of a sawtooth wave

In Figure 7.12, the waveform and frequency spectrum of a 50 Hz sawtooth wave is plotted. A sawtooth wave is comprised of harmonics at frequencies that are odd and even multiples of the fundamental frequency ($100 = 50 \cdot 2, 150 = 50 \cdot 3, 200 = 50 \cdot 4$, etc.). The amplitude of the harmonics decreases with increasing frequency.

Sawtooth Wave Additive Synthesis

An approximation of a sawtooth wave can be created by combining multiple individual harmonics (i.e., sine functions). The following equation can be used to describe a sawtooth

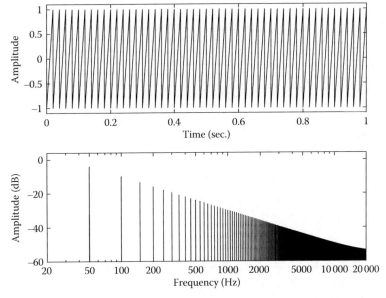

Figure 7.12: Waveform and frequency spectrum of a sawtooth wave

wave as the summation of the odd harmonics of the fundamental frequency:

$$\text{Let: } x = 2 \cdot \pi \cdot f \cdot t$$

$$M = \left\lfloor \frac{F_s}{2 \cdot f} \right\rfloor$$

$$sawtooth(x) = \frac{1}{2} - \frac{1}{\pi}\left(sin(x) + \frac{1}{2}sin(2 \cdot x) + \frac{1}{3}sin(3 \cdot x) + \cdots\right)$$

$$sawtooth(x) = \frac{1}{2} - \frac{1}{\pi}\sum_{n=1}^{M}\frac{1}{n}sin(n \cdot x)$$

7.3.5 Triangle Wave

By default, the amplitude of a sawtooth wave increases linearly from the minimum value to the maximum value. Upon reaching the maximum value, the amplitude immediately returns to the minimum to begin a new cycle. However, upon reaching the maximum value, the amplitude could decrease linearly until it reached the minimum value. The **width** of a sawtooth wave is defined as the proportion of one cycle in which the amplitude is increasing. The following syntax can be used to set the width for a synthesized sawtooth wave: `x = sawtooth(t,width)`. The input variable, `width`, is a number between [0,1]. A `width` value of 0.5 creates a triangle wave with equal proportion for increasing and decreasing

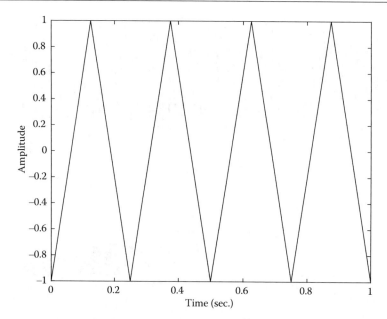

Figure 7.13: Waveform of a triangle wave

amplitude in each cycle. In Example 7.7, a method to synthesize a triangle wave is demonstrated. The result is plotted in Figure 7.13.

Examples: Create and run the following m-file in MATLAB.

```
% TRIANGLESYNTHESIS
% This script demonstrates a method
% to transform the "sawtooth" function to
% a triangle wave
%
% See also SINESYNTHESIS, SAWTOOTHSYNTHESIS, SQUARESYNTHESIS

clear;clc;
% Example - Triangle Wave Signal
% 4 Hz signal for visualization
f = 4; phi = 0; Fs = 40;
Ts = 1/Fs;
t = [0:Ts:1].'; % Time vector in seconds

% Triangle wave => peak occurs at half (0.5) of cycle length
triangleWave = sawtooth(2*pi*f.*t + phi,0.5);
plot(t,triangleWave);
```

```
% Sawtooth with peak at beginning of cycle, then decreasing amp
sawtoothWave = sawtooth(2*pi*f.*t + phi,0);
plot(t,sawtoothWave);

% Sawtooth with increasing amp during cycle, peak at end
sawtoothWave = sawtooth(2*pi*f.*t + phi,1);
plot(t,sawtoothWave);

% 880 Hz signal for audition
Fs = 44100; Ts = (1/Fs);
t = 0:Ts:3; t = t(:);
f = 880; % Frequency in Hz

tr = sawtooth(2*pi*f.*t,0.5); % Triangle
sound(tr,Fs);
```

Example 7.7: Script: `triangleSynthesis.m`

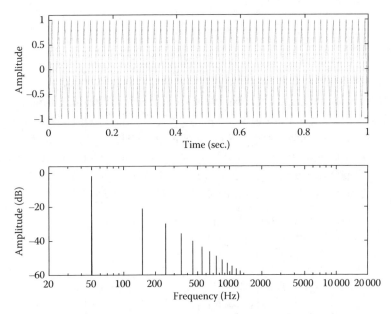

Figure 7.14: Waveform and frequency spectrum of a triangle wave

In Figure 7.14, the waveform and frequency spectrum of a 50 Hz triangle wave is plotted. A triangle wave is comprised of harmonics at frequencies that are odd multiples of the fundamental frequency ($150 = 50 \cdot 3, 250 = 50 \cdot 5, 350 = 50 \cdot 7$, etc.), similar to a square wave (Section 7.3.3). However, the amplitude of the harmonics decreases with a greater slope for the triangle wave than the square wave.

Triangle Wave Additive Synthesis

An approximation of a sawtooth wave can be created by combining multiple individual harmonics (i.e., sine functions). The following equation can be used to describe a square wave as the summation of the odd harmonics of the fundamental frequency:

$$\text{Let: } x = 2 \cdot \pi \cdot f \cdot t$$

$$M = \left\lfloor \frac{F_s}{2 \cdot f} \right\rfloor$$

$$triangle(x) = \frac{8}{\pi^2} \left(sin(x) + \frac{1}{9}sin(3 \cdot x) + \frac{1}{25}sin(5 \cdot x) + \cdots \right)$$

$$triangle(x) = \frac{8}{\pi^2} \sum_{n=1,3,5,\ldots}^{M} \frac{1}{n^2} sin(n \cdot x)$$

Therefore, the decreasing slope for the amplitude of each harmonic for the triangle wave is dictated by $\frac{1}{n^2}$, whereas the square wave is $\frac{1}{n}$. This confirms the visual differences in the frequency spectrum for each signal.

7.3.6 Impulse Train

A digital impulse is a signal with an amplitude of 1 at a single time sample, and an amplitude of 0 at all other time samples. An **impulse train** is created by repeating digital impulses sequentially. The repetition of impulses creates a periodic signal with a cycle defined as the length of time between impulses. There is not a built-in MATLAB function for synthesizing an impulse train. However, it can be created by using the zeros function and changing the amplitude value of specific samples to one. In Example 7.8, a method to synthesize an impulse train is demonstrated. The result is plotted in Figure 7.15.

Examples: Create and run the following m-file in MATLAB.

```
% IMPULSETRAIN
% This script demonstrates a method to create
% an impulse train signal. Initially, all values
% of the signal are set to zero. Then, individual
% samples are changed to a value of 1 based on the
% length of a cycle's period.
clear;clc;

% Example - Impulse Train Signal
```

```
% 5 Hz signal for visualization
f = 5;
Fs = 20; Ts = (1/Fs);
t = 0:Ts:1; t = t(:); % Time vector

impTrain = zeros(size(t)); % Initialize to all zeros
period = round(Fs/f); % # of samples/cycle
% Change the single sample at the start of a cycle to 1
impTrain(1:period:end) = 1;
stem(t,impTrain); axis([0 1 -0.1 1.1]); % Show stem plot

% Example - 440 Hz signal for audition
f = 440;
Fs = 48000; Ts = (1/Fs);
t = 0:Ts:3; t = t(:);
it = zeros(size(t));
period = round(Fs/f); % # of samples/cycle
it(1:period:end) = 1;
sound(it,Fs);   % Listen to signal

% 50 Hz signal for spectrum plot
f = 50;
Fs = 48000; Ts = (1/Fs);
t = 0:Ts:1; t = t(:);
it = zeros(size(t));
period = round(Fs/f); % # of samples/cycle
it(1:period:end) = 1;
plottf(it,Fs);   % Plot spectrum
```

Example 7.8: Script: impulseTrain.m

In Figure 7.16, the waveform and frequency spectrum of a 50 Hz impulse train is plotted. A impulse train is comprised of harmonics at frequencies that are odd and even multiples of the fundamental frequency ($100 = 50 \cdot 2, 150 = 50 \cdot 3, 200 = 50 \cdot 4$, etc.). The amplitude of the harmonics is equal across the spectrum.

Impulse Train Additive Synthesis

The summation of harmonically related sine wave signals at an equal amplitude provides a primative approximation of an impulse train signal. However, it does not create a strictly

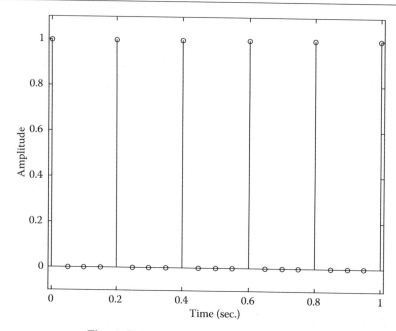

Figure 7.15: Waveform of an impulse train

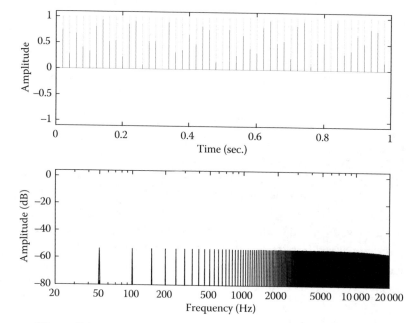

Figure 7.16: Waveform and frequency spectrum of an impulse train

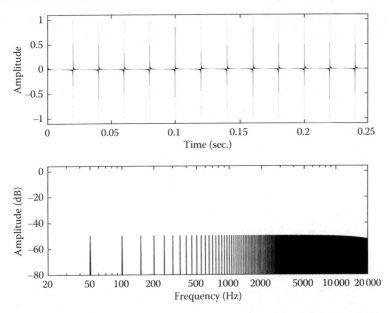

Figure 7.17: Approximation of an impulse train using additive synthesis

positive function. Instead, the waveform shows both positive and negative values near the
impulse. Nonetheless, the frequency spectrum does have harmonics at equal amplitude across
the spectrum, as shown in Figure 7.17.

$$\text{Let: } x = 2 \cdot \pi \cdot f \cdot t$$

$$M = \left\lfloor \frac{F_s}{2 \cdot f} \right\rfloor$$

$$impulseTrain(x) = \frac{\pi}{2 \cdot M} \Big(sin(x) + sin(2 \cdot x) + sin(3 \cdot x) + \cdots \Big)$$

$$impulseTrain(x) = \frac{\pi}{2 \cdot M} \sum_{n=1}^{M} sin(n \cdot x)$$

7.4 Aperiodic Signals

Besides signals based on repeating a cyclical pattern over a period of time, there are also other
audible signals that do not repeat. Therefore, these signals do not have a period, and are called
aperiodic. One example of an aperiodic signal is white noise.

7.4.1 White Noise

White noise is a random signal with the likelihood of equal amplitude at all frequencies.
A white noise signal (of infinite length) has a frequency spectrum with a constant, flat

amplitude. It can be synthesized in MATLAB by using a Gaussian random number generator. The following function can be used to create white noise: randn(n,1).

Examples: Save and run the following m-file in MATLAB.

```
% This script synthesizes a white noise signal by using a
% Gaussian random number generator. Gaussian random numbers
% are also described as normally distributed random numbers.
% The function "randn" creates random numbers based on a
% normal distribution.

Fs = 44100;         % Sampling rate
sec = 5;            % Desired length in seconds
samples = Fs*sec;   % Convert length to samples

% Next, synthesize noise. The scalar (0.2) is to reduce the
% amplitude of the signal to within the full-scale range.
% An additional option would be to perform peak normalization
% to ensure the amplitude is always between -1 to 1.
noise = 0.2*randn(samples,1);
sound(noise,Fs);    % Listen to result
```

Example 7.9: Synthesizing white noise

In Figure 7.18, the waveform and frequency spectrum of a white noise signal is plotted. Because it is an aperiodic signal, the frequency spectrum does not show the presence of

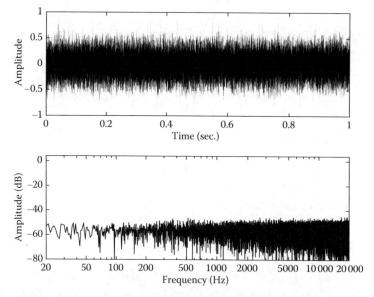

Figure 7.18: Waveform and frequency spectrum of white noise

distinct harmonics (which is a characteristic of periodic signals). Instead, it shows the presense of amplitude at frequencies across the spectrum.

White noise is not the only type of aperiodic signal. Another type of noise used in audio is pink noise, which does not have a flat spectrum. The process of synthesizing pink noise is presented in Sections 12.9 and 13.9.

Bibliography

Aryal, S. (2010, March 16). Example: Unit Circle. Retrieved from www.texample.net/tikz/examples/unit-circle

Chapter 8

Digital Summing, Signal Fades, and Amplitude Modulation
Element-Wise Processing: Arrays Operations with Arrays

8.1 Introduction: Combining Signals

Several fundamental ways to process digital signals involve combining two or more arrays together using an operator. This is an extension of the processes discussed in Chapter 6 using a scalar with an array. By combining arrays together with different operators, it is possible to perform digital summing, ring modulation, amplitude fades, and amplitude modulation.

8.1.1 Array Operations with Arrays

Arrays can be added, subtracted, multiplied, divided, and raised-to-the-power of other arrays. These operations are based on element-wise processing. One requirement for all element-wise operations is the two arrays be the same dimensions. Element-wise multiplication, division, and power operators require a period (".") prior to the operator to indicate the element-wise or **point-wise** processing. In this case, the element of each index of one array is applied to the element of the same index of the other array.

Examples: Execute the following commands in MATLAB.

```
>> x = [ 1 2 3 ];
>> y = [ 2 3 2 ];
>> addVec = x + y
>> z = [ 2 ; 3 ; 2 ];
>> addErr = x + z
>> x + z.'
>> subVec = x - y
>> multVec = x .* y
```

```
>> divVec = x ./ y
>> powVec = x .^ y
>> w = [ 1 2 3 ; 4 5 6 ];
>> v = [ 1 2 1 ; 2 1 2 ];
>> addMat = v + w
>> subMat = v - w
>> multMat = v .* w
>> divMat = v ./ w
>> powMat = v .^ w
```

Example 8.1: Element-wise array operations

8.1.2 Using Array Operations to Combine Signals

The operations with arrays demonstrated in Example 8.1 are the basis for how signals can be combined together. This process involves using two or more signals to create a new signal by combining all of the signals. There are primarily two operations which can be used to perform several processes. First, array addition is used for digital summing to perform the task of a mixing console. Second, array multiplication is used for ring modulation and signal fades, as well as amplitude modulation for the tremolo effect.

8.2 Array Addition: Signal Summing

Signals can be combined by adding together the amplitude of each signal. For digital signals, this involves adding the amplitude at one time sample for one signal with the amplitude at the same time sample for the other signal(s). Therefore, the addition of digital signals in MATLAB is accomplished by an element-wise addition of arrays. Blending or mixing multiple signals together through this process is a fundamental task of digital audio workstation software.

8.2.1 Addition Block Diagram

A basic block diagram, shown in Figure 8.1, displays the input signals, x and w, routed to the output signal, y. This process can be written mathematically: $y[n] = x[n] + w[n]$. For signals $x = \{x_1, x_2, ..., x_N\}$ and $w = \{w_1, w_2, ..., w_N\}$, the command y = x + w assigns the addition of

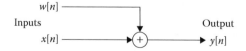

Figure 8.1: Block diagram of digital summing

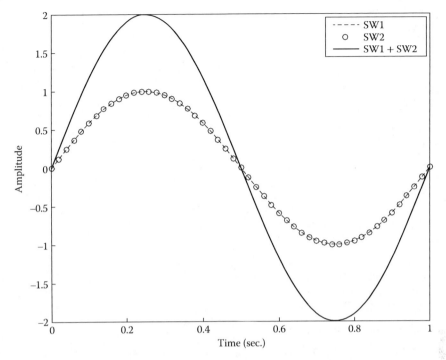

Figure 8.2: Adding sine waves with identical frequency and phase

each individual element in x and w to the corresponding element of y, such that $y_1 = x_1 + w_1$, $y_2 = x_2 + w_2, \ldots , y_N = x_N + w_N$. In Example 8.2, two methods are demonstrated to add signals together. The results are plotted in Figures 8.2 and 8.3.

Examples: Create and run the following m-file in MATLAB.

```
% ADDITIONEXAMPLE
% This script provides two examples for combining
% signals together using addition
%
% This first example is for signals of the
% same frequency and phase
%
% The second example is for signals of different frequencies
%
% See also SUBTRACTIONEXAMPLE

% Example 1 - Same Frequencies
% Declare initial parameters
```

```matlab
f = 1;
a = 1;
phi = 0;
Fs = 100;
t = [0:1/Fs:1].';
sw1 = a * sin(2*pi*f.*t+phi);
sw2 = a * sin(2*pi*f.*t+phi);
N = length(sw1);
sw3 = zeros(N,1);
% Loop through arrays to perform element-wise addition
for n = 1:N
    sw3(n,1) = sw1(n,1)+sw2(n,1);
end
% Plot the result
plot(t,sw1,'--',t,sw2,'o',t,sw3);
xlabel('Time (sec.)');
ylabel('Amplitude');
title('Addition of 2 Sine Waves - Same Frequencies');
legend('SW1','SW2','SW1+SW2');

% Example 2 - Different Frequencies
% Declare initial parameters
f = 1;
a = 1;
phi = 0;
Fs = 100;
t = 0:1/Fs:1;
sw1 = a * sin(2*pi*f.*t+phi);
sw2 = a * sin(2*pi*(f*2).*t+phi); % Change frequency x2
% Addition of arrays in MATLAB is element-wise by default
sw3 = sw1+sw2;

figure; % Create a new figure window
plot(t,sw1,'--',t,sw2,'.',t,sw3);
xlabel('Time (sec.)');
ylabel('Amplitude');
title('Addition of 2 Sine Waves - Different Frequencies');
legend('SW1','SW2','SW1+SW2');
```

Example 8.2: Script: `additionExample.m`

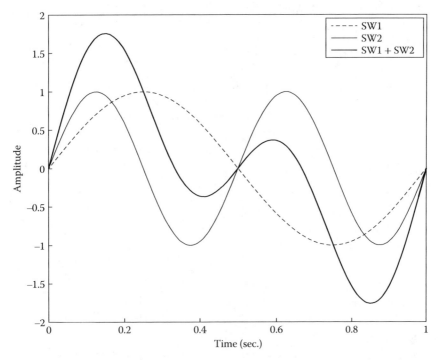

Figure 8.3: Adding sine waves with different frequencies

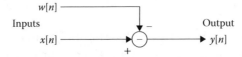

Figure 8.4: Block diagram of signal subtraction

8.2.2 Array Subtraction

A second method of combining signals is using element-wise subtraction. In general, signal subtraction is the equivalent of signal addition after inverting the polarity of the second signal. Conceptually, this method of combining signals results in destructive interference for signals of identical frequency and phase. Audio engineers use this type of processing in an analysis called the **null test** to determine if two signals are identical.

Subtraction Block Diagram

A block diagram, shown in Figure 8.4, displays the subtraction of input signals, x and w, routed to the output signal, y. For signals $x = \{x_1, x_2, ..., x_N\}$ and $w = \{w_1, w_2, ..., w_N\}$, the command y = x - w assigns the result of subtracting each individual element in w from x to the corresponding element of y, such that $y_1 = x_1 - w_1$, $y_2 = x_2 - w_2$, ... , $y_N = x_N - w_N$. In

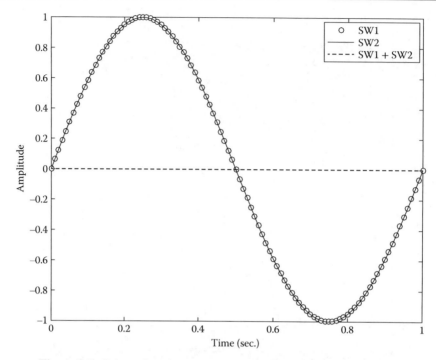

Figure 8.5: Subtracting sine waves with identical frequency and phase

general, this process can be written $y[n] = x[n] - w[n]$. In Example 8.3, two methods are demonstrated to subtract one signal from another. The results are plotted in Figures 8.5 and 8.6.

Examples: Create and run the following m-file in MATLAB.

```
% SUBTRACTIONEXAMPLE
% This script provides two examples for combining signals
% together using subtraction
%
% This first example is for signals of the same frequency and phase
%
% The second example shows the addition of signals where
% one signal has a phase offset of 180 degrees (pi radians)
%
% See also ADDITIONEXAMPLE


% Example 1 - Same Frequency and Phase
```

```
% Declare initial parameters
f = 1;  a = 1; phi = 0; Fs = 100;
t = 0:1/Fs:1;
sw1 = a * sin(2*pi*f.*t+phi); sw2 = a * sin(2*pi*f.*t+phi);
% Element-wise subtraction
sw3 = sw1 - sw2;
% Plot the result
plot(t,sw1,'.',t,sw2,t,sw3,'--');
xlabel('Time (sec.)');
ylabel('Amplitude');
title('Subtraction of 2 Sine Waves - Same Frequency and Phase');
legend('SW1','SW2','SW1-SW2');


% Example 2 - Same frequency with a phase offset

sw1 = a * sin(2*pi*f.*t+0);
sw2 = a * sin(2*pi*f*t+pi); % Phase offset by 180 degrees

sw3 = sw1+sw2;

figure; % Create a new figure window
plot(t,sw1,'.',t,sw2,t,sw3,'--');
xlabel('Time (sec.)');
ylabel('Amplitude');
title('Addition of 2 Sine Waves - 180 degree phase offset');
legend('SW1','SW2','SW1-SW2');
```

Example 8.3: Script: subtractionExample.m

8.3 Array Multiplication: Ring Modulation

Another operation that is used with arrays is multiplication. For signals, $x = \{x_1, x_2, ..., x_N\}$ and $w = \{w_1, w_2, ..., w_N\}$, the command y=x.*w assigns the multiplication of each individual element in x and w to the corresponding element of y, such that $y_1 = x_1 \cdot w_1$, $y_2 = x_2 \cdot w_2$, ... , $y_N = x_N \cdot w_N$. Mathematically, this process can be written $y[n] = x[n] \cdot w[n]$.

One application of array multiplication is **ring modulation**. In this process, the amplitude values of two signals are multiplied together. It is typical for at least one of the signals to be a basic, periodic signal like a sine wave.

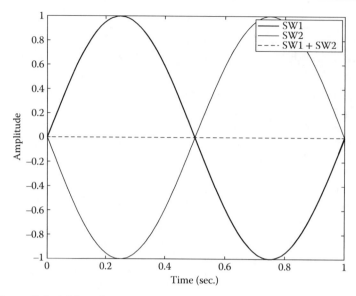

Figure 8.6: Adding sine waves with the same frequency and a 180° offset

The result of performing ring modulation can be understood by evaluating the multiplication of two sine wave signals. First, consider a condition when the frequency, f, is identical in both sine waves. By using a trigonometric identity, the result is a signal with twice the original frequency, half the amplitude, a DC offset of $\frac{1}{2}$, and a phase shift of $-\frac{\pi}{2}$.

$$sin^2(\theta) = \frac{1 - cos(2\theta)}{2}$$

$$sin(2\pi \cdot f \cdot t) \cdot sin(2\pi \cdot f \cdot t) = \frac{1}{2} - \frac{cos(2\pi \cdot (2f) \cdot t)}{2}$$

$$= \frac{1}{2} \cdot sin(2\pi \cdot (2f) \cdot t - \frac{\pi}{2}) + \frac{1}{2}$$

Next, consider the condition when one sine wave has a higher frequency, f_H, than the second sine wave's frequency, f_L, such that $f_H > f_L$. The combined signal has two frequencies, $f_1 = f_H - f_L$ and $f_2 = f_H + f_L$. Using another trigonometric identity:

$$sin(\theta) \cdot sin(\phi) = \frac{cos(\theta - \phi) - cos(\theta + \phi)}{2}$$

$$sin(2\pi \cdot f_H \cdot t) \cdot sin(2\pi \cdot f_L \cdot t) = \frac{cos(2\pi \cdot (f_H - f_L) \cdot t) - cos(2\pi \cdot (f_H + f_L) \cdot t)}{2}$$

$$= \frac{1}{2} \cdot \left(sin(2\pi \cdot (f_H - f_L) \cdot t + \frac{\pi}{2}) + sin(2\pi \cdot (f_H + f_L) \cdot t - \frac{\pi}{2}) \right)$$

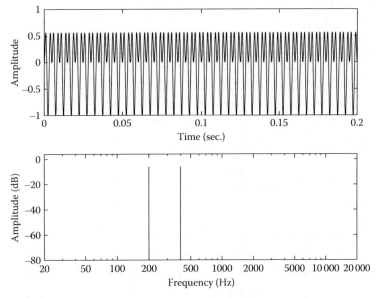

Figure 8.7: Waveform and frequency spectrum of signal synthesized by ring modulation

In Example 8.4, ring modulation is performed with two sine waves with frequencies of 300 Hz and 100 Hz, respectively. The result is a signal with harmonics of 200 Hz and 400 Hz, shown in Figure 8.7.

Examples: Create and run the following m-file in MATLAB.

```
% RINGMODULATION
%
% This script demonstrates the result of performing
% ring modulation with two different frequencies.
%
% By multiplying 300 Hz with 100 Hz, the result
% is a signal with two harmonics:
% 200 Hz (300 - 100) and 400 Hz (300 + 100)

clc; clear;
% Initial parameters
Fs = 48000; Ts = 1/Fs;
lenSec = 2;
N = lenSec * Fs;
fHigh = 300;
fLow = 100;
```

```
% Synthesize signals and perform element-wise multiplication
x = zeros(N,1);
for n = 1:N
    t = (n-1) * Ts;
    x(n,1) = sin(2*pi*fLow*t) * sin(2*pi*fHigh*t);
end


plottf(x,Fs);
```

Example 8.4: Script: `ringModulation.m`

An interesting consequence occurs when three sine wave signals are multiplied together with frequencies, $f_L < f_M < f_H$. Although there are three signals with one frequency each being combined together, the result is a signal that can have four harmonics. The frequencies of those harmonics are:

$$f_H - (f_M + f_L)$$
$$f_H - (f_M - f_L)$$
$$f_H + (f_M - f_L)$$
$$f_H + (f_M + f_L)$$

In Figure 8.8, the result of performing ring modulation with three sine waves is shown. For this example, the original frequencies of the sine waves are 100, 300, and 1000 Hz. The combined signal has harmonics at 600, 800, 1200, and 1400 Hz.

Furthermore, performing ring modulation with four sine waves produces a signal that can have eight harmonics. In general, performing ring modulation with n sine waves produces a signal which can have $2^{(n-1)}$ harmonics.

8.4 Array Multiplication: Amplitude Fade

Another application of array multiplication is to **fade** the amplitude of a signal. A *fade-in* can be used at the beginning of a signal to gradually increase the amplitude of a signal from 0 to unity gain. A *fade-out* can be used at the end of a signal to gradually decrease the amplitude of a signal from unity gain to 0. There are several common types of fades including linear, exponential, and S-curve.

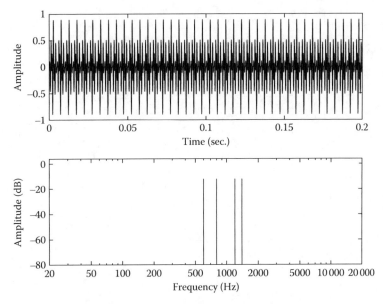

Figure 8.8: Ring modulation with three sine waves

8.4.1 *Linear Fade*

A **linear fade** changes the amplitude of a signal based on a function with a constant slope. This function can be created by using the following MATLAB function:

$$linspace(0,1,numOfSamples).$$

Examples: Create and run the following m-file in MATLAB.

```
% LINEARFADE
% This script creates linear fades. Then,
% the "fade-in" is applied to the beginning of
% a sine wave signal. The "fade-out" is applied
% to the end.
%
% See also EXPONENTIALFADE

clc;clear;close all;
Fs = 48000;
Ts = (1/Fs);
t = 0:Ts:3;
t = t(:);
f = 100; phi = 0;
in = sin(2*pi*f.*t + phi);
figure(1);
```

```
plot(t,in);

numOfSamples = 1*Fs; % 1 second fade-in/out
a = linspace(0,1,numOfSamples); a = a(:);
fadeIn = a;
fadeOut = 1-a; % Equivalent = linspace(1,0,numOfSamples);
figure(2);
plot(a,fadeIn,a,fadeOut);legend('Fade-in','Fade-out');

% Fade-in
% Index samples just at the start of the signal
temp = in;
temp(1:numOfSamples) = fadeIn .* in(1:numOfSamples);
figure(3);
plot(t,temp);

% Fade-out
% Index samples just at the end of the signal
out = temp;
out(end-numOfSamples+1:end) = fadeOut.*temp(end-numOfSamples+1:end);
figure(4);
plot(t,out);
```

Example 8.5: Script: linearFade.m

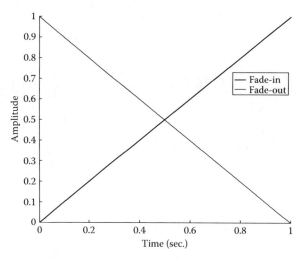

Figure 8.9: Linear fade curves

In Example 8.5, a method to create a linear fade is demonstrated. In Figure 8.9, the linear fade curves are shown. The result of applying the linear fades to a signal is shown in Figure 8.10.

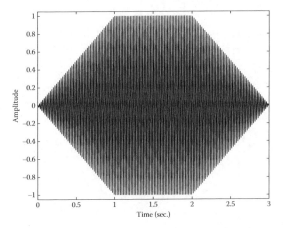

Figure 8.10: Linear fade applied to signal

8.4.2 Exponential Fade

An **exponential fade** can be used to increase or decrease the amplitude of a signal based on a curved, exponential function. There are two general types of exponential fades: concave and convex. In both types, the shape of the curve can be variably changed based on the value of an exponent.

In Example 8.6, a method to create an exponential fade is demonstrated. In Figure 8.11, the convex exponential fade curves are shown. The result of applying the convex exponential fades to a signal is shown in Figure 8.12. Additionally, concave exponential fade curves are shown in Figure 8.13.

Examples: Create and run the following m-file in MATLAB.

```
% EXPONENTIALFADE
% This script creates exponential fades. Both
% convex and concave examples are provided.
% These fades are applied to the beginning and end
% of a sine wave test signal.
%
% See also LINEARFADE

clear;clc;close all;
Fs = 48000;
Ts = (1/Fs);
t = 0:Ts:3;
t = t(:);
f = 100; phi = 0;
```

```
in = sin(2*pi*f.*t + phi);
figure(1);
plot(t,in);

% Convex fades
numOfSamples = 1*Fs; % 1 second fade-in/out

% Exponent for curve
x = 2; % "x" can be any number > 1 (linear  = 1)

a = linspace(0,1,numOfSamples); a = a(:);
fadeOut = 1 - a.^x;

a = linspace(1,0,numOfSamples); a = a(:);
fadeIn = 1 - a.^x;

figure(2);
plot(a,fadeIn,a,fadeOut);legend('Fade-in','Fade-out');

% Fade-in
temp = in;
temp(1:numOfSamples) = fadeIn .* in(1:numOfSamples);
figure(3);
plot(t,temp);

% Fade-out
out = temp;
out(end-numOfSamples+1:end) = fadeOut.*temp(end-numOfSamples+1:end);
figure(4);
plot(t,out);

% (Alternate) concave fades
a = linspace(0,1,numOfSamples); a = a(:);
fadeOut = a.^x;
a = linspace(1,0,numOfSamples); a = a(:);
fadeIn = a.^x;
figure(5);
plot(a,fadeIn,a,fadeOut);legend('Fade In','Fade Out');
```

Example 8.6: Script: exponentialFade.m

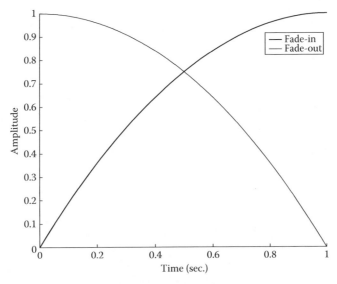

Figure 8.11: Convex exponential fade curves

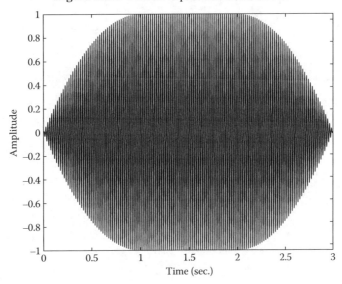

Figure 8.12: Convex exponential fade applied to signal

8.4.3 S-Curve Fade

A third type of amplitude fade is the **S-curve**, named for the shape of the function's ramp. When the function is close to zero, a concave curve is used for the fade. When the function is close to one, a convex curve is used for the fade. One approach to creating the S-curve fade is to concatenate separate concave and convex curves together. In Example 8.7, a method to create an S-curve fade is demonstrated. In Figure 8.14, the S-curve fades are shown.

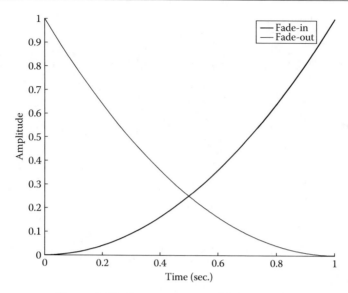

Figure 8.13: Concave exponential fade curves

Examples: Create and run the following m-file in MATLAB.

```
% SCURVEFADE
% This script demonstrates one approach to creating
% S-curve fades. This method involves concatenating
% a convex fade with a concave fade.
%
% See also LINEARFADE, EXPONENTIALFADE, SINECURVEFADE


clear;clc;


Fs = 44100; % Arbitrary sampling rate


% S-curve fade-in
numOfSamples = round(1*Fs); % 1 sec. fade, round to whole sample
a = linspace(0,1,numOfSamples/2); a = a(:);
x = 2; % x can be any number ≥ 1
concave = 0.5*(a).^x;
convex = 0.5*(1 - (1-a).^x)+0.5;
fadeIn = [concave ; convex];


% S-curve fade-out
x = 3; % x can be any number ≥ 1
```

```
convex = 0.5*(1 - a.^x)+0.5;
concave = 0.5*(1-a).^x;
fadeOut = [convex ; concave];


% Plot the S-curves
t = linspace(0,1,numOfSamples); t=t(:);
plot(t,fadeIn,t,fadeOut); legend('Fade-in','Fade-out');
```

Example 8.7: Script: sCurveFade.m

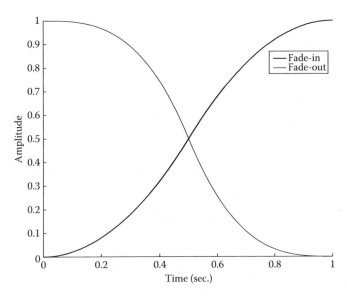

Figure 8.14: S-curve fade

Sine Function S-Curve

An alternative method to creating the S-curve fade is based on the sine function. The fade is created by synthesizing half of a cycle of the sine function, which lasts for the desired length of the fade.

When a sine function is synthesized, it has a peak-to-peak amplitude over a range of −1 to 1. For an amplitude fade, the function should fit within a range of 0 to 1. If the amplitude fade has a negative value, not only is the amplitude scaled but the polarity is inverted, which is undesirable. Therefore, the amplitude of the sine function must be scaled and offset.

$$fadeIn = 0.5 \cdot sin(2 \cdot \pi \cdot f \cdot t - \frac{\pi}{2}) + 0.5$$

$$fadeOut = 0.5 \cdot sin(2 \cdot \pi \cdot f \cdot t + \frac{\pi}{2}) + 0.5$$

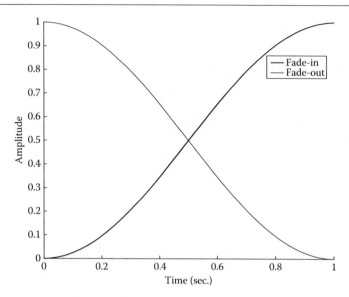

Figure 8.15: Sine function S-curve fade

In Example 8.8, a method to create an S-curve fade using the sine function is demonstrated. In Figure 8.15, the S-curve fades are shown.

Examples: Create and run the following m-file in MATLAB.

```
% SINECURVEFADE
% This script demonstrates one approach to creating
% S-curve fades. This method involves using
% the sine function.
%
% See also LINEARFADE, EXPONENTIALFADE. SCURVEFADE

clear;clc;

Fs = 44100; % Arbitrary sampling rate
Ts = 1/Fs;
% S-curve fade-in
lenSec = 1; % 1 second fade-in/out
N = round(lenSec * Fs);   % Convert to whole # of samples
t = [0:N-1]*Ts; t = t(:);

% The S-curve fade is half a cycle of a sine wave
% If fade is 1 sec.. period of sine wave is 2 sec.
```

```
period = 2 * N * Ts;    % units of seconds
freq = 1/period;        % units of Hz
fadeIn = 0.5*sin(2*pi*freq*t - pi/2) + 0.5;


% S-curve fade-out
fadeOut = 0.5*sin(2*pi*freq*t + pi/2) + 0.5;


% Plot the S-curves
plot(t,fadeIn,t,fadeOut); legend('Fade-in','Fade-out');
```

Example 8.8: Script: `sineCurveFade.m`

8.4.4 Equal-Amplitude and Equal-Power Fades

An important characteristic audio engineers use for selecting one of the fade types is whether a fade creates an **equal-amplitude** or **equal-power** crossfade. These characteristics change how a crossfade between two signals is perceived by a listener.

The combined amplitude for a crossfade is calculated by adding together the combined amplitude of each fade at each time sample ($A_c[n] = A_{in}[n] + A_{out}[n]$). For an equal-amplitude crossfade, the combined amplitude between the two fades always adds together to be unity gain. The equal-amplitude crossfade ensures that if two signals separately have an amplitude bounded within the full-scale range, $[-1, 1]$, then using a crossfade produces a new signal within the full-scale range (i.e., does not clip). However, the drawback of an equal-amplitude crossfade is listeners perceive the combined signal strength to be less in the middle of the fade than the strength at the beginning and end of the fade.

Conversely, an equal-power crossfade has the benefit that listeners perceive the combined signal strength in the middle of the fade to be equal to the signal strength at the beginning and end of the fade. The drawback to the equal-power crossfade is the combined amplitude in the middle of the fade is greater than 1, making it possible for the combined amplitude to be greater than 1 (i.e., clipping). The combined power for a crossfade is calculated by adding together the amplitude squared of the fade-in and fade-out at each time sample, $P_c[n] = (A_{in}[n])^2 + (A_{out}[n])^2$.

In Example 8.9 and Figure 8.16, a comparison is provided between the amplitude and power of a crossfade.

Examples: Create and run the following m-file in MATLAB.

```
% EQUALFADES
% This script analyzes an exponential crossfade
% for "equal amplitude" or "equal power"
```

```
clear;clc;close all;
Fs = 44100;

% Square-root fades
x = 2; % x can be any number ≥ 2
numOfSamples = 1*Fs; % 1 second fade-in/out
aIn = linspace(0,1,numOfSamples); aIn = aIn(:);
fadeIn = (aIn).^(1/x);


aOut = linspace(1,0,numOfSamples); aOut = aOut(:);
fadeOut = (aOut).^(1/x);

% Compare amplitude vs. power of crossfade
plot(aIn,fadeIn,aIn,fadeOut,aIn,fadeIn+fadeOut,...
    aIn,(fadeIn.^2) + (fadeOut.^2));
axis([0 1 0 1.5]);
legend('Fade-in','Fade-out','Crossfade Amplitude','Crossfade Power');
```

Example 8.9: Script: `equalFades.m`

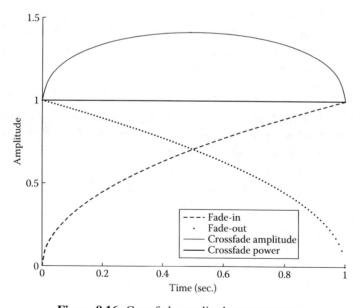

Figure 8.16: Crossfade amplitude versus power

8.5 *Array Multiplication: Amplitude Modulation*

Amplitude modulation (AM) is a process of changing the amplitude of a signal based on the amplitude of another signal. AM is accomplished using element-wise multiplication between two arrays. It is similar to an amplitude fade because AM involves changing the amplitude of a signal between a value of 0 and unity gain. It is also related to ring modulation because a periodic signal is typically used in the AM process.

8.5.1 *Amplitude Modulation Block Diagram*

A block diagram, shown in Figure 8.17, displays the modulation of an input (carrier) signal, x, by the modulator signal, w. The processing represented by this block diagram can be accomplished using the MATLAB command y=x.*w.

8.5.2 *Tremolo*

Tremolo is another name for amplitude modulation when used as an audio effect. For the tremolo effect, the amplitude of a carrier signal is modulated by a low-frequency oscillator (LFO). One type of LFO is a sine wave signal with a frequency less than 20 Hz. Other signals such as a square wave, triangle wave, and sawtooth wave can also be used.

As with amplitude fades, the purpose of amplitude modulation is only to reduce the amplitude of the original signal. This is accomplished by multiplying the signal by a value between unity gain (no reduction) and 0 (no amplitude). It is not desirable to change the polarity of the carrier signal by multiplying the amplitude of a carrier signal by a modulator with negative amplitude. Therefore, the peak-to-peak amplitude of the modulator should be limited to a range of [0,1].

A sine wave signal, $sin(2 \cdot \pi \cdot f \cdot t + \phi)$, should be transformed prior to use as a LFO modulator to ensure the peak-to-peak amplitude is bound by the range [0,1]. One example would be to decrease the amplitude of the sine wave by 0.5 and then apply a constant DC offset of 0.5, $A[n] = 0.5 \cdot sin(2 \cdot \pi \cdot f \cdot t + \phi) + 0.5$. Other transformations can also be used if they ensure the modulator amplitude does not fall outside the range of [0,1]. A general method of transforming linear scales is presented in the appendix for this chapter (Section 8.6).

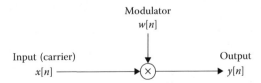

Figure 8.17: Block diagram of amplitude modulation

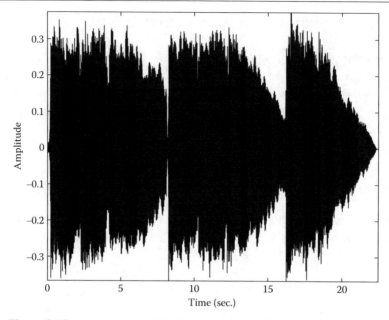

Figure 8.18: Amplitude modulation: unprocessed rhythm guitar (carrier)

There are several typical parameters of a tremolo effect. The first parameter is the *rate* (or speed) of modulation, which can be changed by increasing or decreasing the frequency of the modulator. The second parameter is the *depth* of modulation, also called the *intensity* of the effect, which can range from 0% to 100%. Depth can be used to change the amplitude and DC offset so the LFO fits with the range [0,1]. If the depth is set to the maximum (100%), an LFO should have a peak-to-peak amplitude of 1 and a DC offset of 0.5. For the minimum depth (0%), the amplitude of the LFO should be 0 with a DC offset of 1.

In Example 8.10, the process of amplitude modulation is demonstrated. In Figure 8.18, the waveform of the original signal is shown. In Figure 8.19, the process of transforming the modulation LFO to the appropriate range is shown. In Figures 8.20 and 8.21, a short segment is plotted of the unprocessed and processed signals, respectively.

Examples: Create and run the following m-file in MATLAB.

```
% AMPMODULATION
% This script provides an example for modulating
% the amplitude of a carrier signal. This process
% is called tremolo when used as an audio effect
```

```
clear;clc;close all;

% Import carrier signal
[carrier,Fs] = audioread('RhythmGuitar.wav');
Ts = 1/Fs;
N = length(carrier);
t = [0:N-1]*Ts; t = t(:);
plot(t,carrier);title('Original Soundfile (Electric Guitar)');
xlabel('Time (sec.)');figure;

% Tremolo parameters
depth = 100; % [0.100]
speed = 5;
amp = 0.5*(depth/100);
offset = 1 - amp;

% Synthesize modulation signal
f = speed;  % speed of effect
phi = 0;
sw = sin(2*pi*f.*t+phi);

mod = amp.*sw + offset;

% Plot to compare the original sine wave with the modulator
% Index only first second of signal for visualization purposes
plot(t(1:Fs),sw(1:Fs),t(1:Fs),mod(1:Fs));
title('Modulator Signal');figure;

% Modulate the amplitude of the carrier by the modulator
output = carrier.*mod;

% Plot the output and listen to the result
plot(t(1:Fs),carrier(1:Fs));
title('Carrier Signal (Unprocessed)');
figure;plot(t(1:44100),output(1:44100));
title('Output Signal (Processed)');
sound(output,Fs);
```

Example 8.10: Script: ampModulation.m

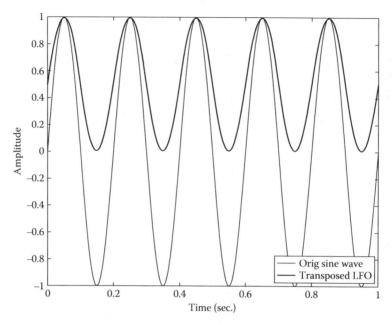

Figure 8.19: Amplitude modulation: sine wave LFO (modulator)

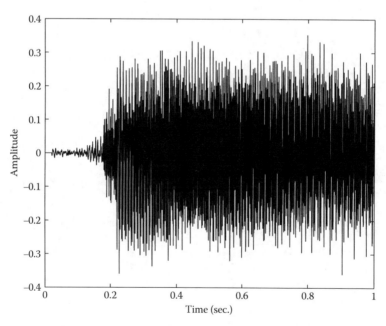

Figure 8.20: Amplitude modulation: original carrier (unprocessed)

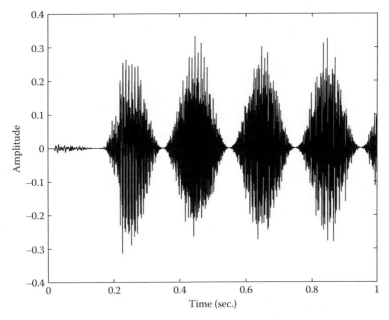

Figure 8.21: Amplitude modulation: output (processed signal)

Morphing the LFO from a Triangle Wave to a Square Wave

On some tremolo effects, a variable parameter is used to morph the LFO from a triangle wave to a square wave. In software code, a synthesized triangle wave signal can be processed using mathematical operations to resemble a square wave. In the following example, it is assumed a LFO variable knob can have values from 1–10. When the knob value is 1, a triangle LFO is synthesized. As the knob value increases to 10, the LFO more closely resembles a square wave. To process the triangle wave, the positive amplitude values are processed separately from the negative amplitude values. In Example 8.11, a method to synthesize an LFO signal that can morph from a triangle wave to a square wave is demonstrated. In Figure 8.22, the waveforms of two example LFOs are shown.

Examples: Create and run the following m-file in MATLAB.

```
% MORPHLFO
%
% This script demonstrates the method of morphing
% the LFO signal for the tremolo effect from
% a triangle wave to a square wave
%
% See also AMPMODULATION
```

```
clear; clc; close all;

% Initial parameters
Fs = 48000; Ts = 1/Fs;
f = 1; % 1 Hz LFO
t = 0:Ts:1; t = t(:);

% Consider a parameter knob with values from 1-10
knobValue = 10;
lfo = sawtooth(2*pi*f*t+pi/2,0.5);
N = length(lfo);
lfoShape = zeros(N,1);
for n = 1:N

    if lfo(n,1) >= 0
        % This process is similar to adding an exponent
        % to a linear fade. It turns it into an
        % exponential, convex curve. In this case, it
        % turns the straight line of a triangle wave
        % into a curve closer to a square wave.
        lfoShape(n,1) = lfo(n,1) ^ (1/knobValue);
    else
        % Need to avoid using negative numbers with power function
        lfoShape(n,1) = -1*abs(lfo(n,1)) ^ (1/knobValue);
    end
end

plot(t,0.5*lfo+0.5,t,0.5*lfoShape+0.5);
legend('Triangle, Knob = 1','Square, Knob = 10');
```

Example 8.11: Script: `morphLFO.m`

8.6 Appendix: Transforming Linear Scales

There are many applications in audio signal processing which require converting a scalar or signal from one linear scale to a different linear scale. The following describes a general

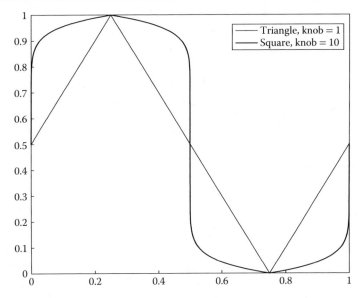

Figure 8.22: Amplitude modulation: morphing the LFO

method for transforming the values from one scale to a new scale. Suppose x has a range, m_o, and offset, μ_o. It desired to transform x to y with a new range, m_n, and new offset, μ_n.

$$\text{Let: } y = \frac{m_n}{m_o} \cdot x - \mu_o + \mu_n$$

$$\text{Where: } m_o = \max(x) - \min(x)$$

$$m_n = \max(y) - \min(y)$$

$$\mu_o = 0.5 \cdot (\max(x) + \min(x))$$

$$\mu_n = 0.5 \cdot (\max(y) + \min(y))$$

Stereo Panning and Mid/Side Processing
Element-Wise Processing: Two-Channel Audio

9.1 Introduction: Stereo Audio Signals

It is common for audio signals to be played on stereo systems consisting of two speakers (i.e., headphones or loudspeakers). A recorded, mono signal from a single microphone is converted to a two-channel, stereo signal to be played through one or both of the speakers at the same time. Audio engineers control the perceived position across the stereo field of a recorded signal using panning. Another processing technique used to add audio effects to a stereo signal is mid/side processing. These topics are presented in the current chapter.

9.2 Stereo Panning

Stereo panning is the method of processing a mono audio signal to be a stereo audio signal. Typically, panning is performed by changing the amplitude of the signal in the left channel and the right channel. Analog consoles use a panning potentiometer to give an engineer variable control over how much signal is sent to the left channel and the right channel. In digital audio, panning is accomplished by processing a single audio signal to duplicate it for the left and right channels. The amplitude of the duplicated versions can be scaled to change how much of the signal's amplitude is sent to the left and the right channels. A block diagram describing the basic process to create a stereo signal is shown in Figure 9.1.

The common convention for panning is for the potentiometer range to span from −100 (left) to +100 (right). For a value of 0 (center) an equal signal level is provided to both channels. When a signal is panned to −100 (left), the amplitude in the left channel is identical to the amplitude of the original signal. However, the amplitude in the right channel from the duplicate signal is reduced to zero. When a signal is panned to values between −100 (left) and 100 (right), the amplitude of the duplicated signal in each channel is scaled by a value between 0 and 1. A block diagram describing the process of scaling the amplitude of the left and right channels of a stereo signal is shown in Figure 9.2.

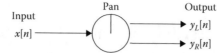

Figure 9.1: Block diagram of mono to stereo process

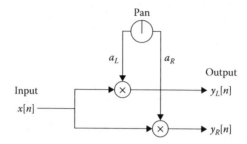

Figure 9.2: Block diagram of panning process

9.2.1 Panning Functions

A **panning function** can be used to determine the amplitude used for the signal in the left channel and the right channel. Panning functions range in amplitude between 0 and 1. There are several different panning functions an audio engineer can choose. In all cases, the function for the left channel complements the function for the right channel to maintain symmetry during panning. In Figure 9.3, three examples of panning functions are plotted for comparison.

In a way, a panning potentiometer performs a type of amplitude fade on stereo signals based on a panning function. In fact, there are close similarities between stereo panning functions and amplitude fades (Section 8.4).

Linear panning uses a function such that when a signal is panned to the center, the amplitude of the original signal is reduced by half in each channel. Linear panning achieves a constant gain, or equal combined amplitude, for all possible panning positions. The disadvantange of linear panning is that signal power is reduced for signals panned to the center. There are other panning functions (square-law panning, sine-law panning) that can be used to maintain signal power for center-panned signals.

The following equations can be used to determine the amplitude scalar applied to each channel for various panning functions. A normalized pan value, p_N, is used in each equation. It is a linear transformation of the panning value, p, from a scale of -100 to 100 to a scale of 0–1, where $p_N = \frac{p}{200} + 0.5$.

$$\text{linear panning} = \begin{cases} a_L = 1 - p_N \\ a_R = p_N \end{cases}$$

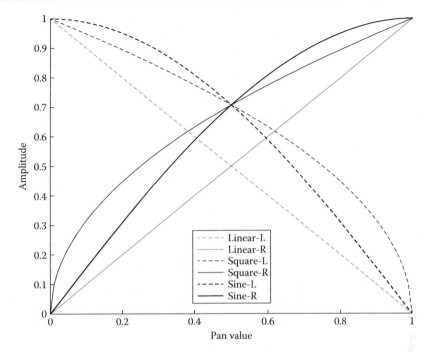

Figure 9.3: Plot of common panning functions

$$\text{square-law panning} = \begin{cases} a_L = \sqrt{1-p_N} \\ a_R = \sqrt{p_N} \end{cases}$$

$$\text{sine-law panning} = \begin{cases} a_L = sin\left((1-p_N)\cdot\frac{\pi}{2}\right) \\ a_R = sin\left(p_N\cdot\frac{\pi}{2}\right) \end{cases}$$

For the special case of center panning, the value of the amplitude scalar is identical for both the left and right channels. For linear panning, the value is 0.5. This type of panning is labeled in some audio software as −6 dB planning ($20\cdot log_{10}(0.5)$). For square-law panning and sine-law panning, the value of the amplitude scalar is ∼0.707. Another description of this type is −3 dB panning ($20\cdot log_{10}(0.707)$).

Examples: Create the following function m-file in MATLAB.

```
% PAN
% This function pans a mono audio signal in a stereo field.
% It is implemented such that it can pan the entire signal
% to one location if "panValue" is a scalar. It can also be
% used for auto-pan effects if "panValue" is an array.
```

```
%
% Input Variables
%    panType : 1 = linear, 2 = sqRt, 3 = sine-law
%    panValue : (-100 to +100) transformed to a scale of (0-1)

function [out] = pan(in, panValue, panType)

% Convert pan value to a normalized scale
panTransform = (panValue/200) + 0.5;

% Conditional statements determining panType
if panType == 1
    leftAmp = 1-panTransform;
    rightAmp = panTransform;
elseif panType == 2
    leftAmp = sqrt(1-panTransform);
    rightAmp = sqrt(panTransform);
elseif panType == 3
    leftAmp = sin((1-panTransform) * (pi/2));
    rightAmp = sin(panTransform * (pi/2));

end

leftChannel = leftAmp.*in;
rightChannel = rightAmp.*in;

out = [leftChannel,rightChannel];

end
```

Example 9.1: Function: `pan.m`

Equal-Amplitude and Equal-Power Panning

An important characteristic audio engineers use for choosing one of the panning functions is whether it has **equal amplitude** or **equal power**. These characteristics change how panning is perceived by a listener. This is similar to the equal-amplitude and equal-power crossfades discussed in Section 8.4.4.

A panning function has equal amplitude when the combined amplitude of the left channel and the right channel is equal for every panning position, $A_c = A_L + A_R$. The linear panning functions fulfill this requirement. A panning function has equal power when the combined power is equal for every panning position, $P_c = (A_L)^2 + (A_R)^2$. The square-law and sine-law panning functions are both equal power.

Ideally, audio engineers use panning to move the perceived location of a sound source across the stereo field. If an audio engineer pans a signal from the center position over to the left side, the perceived strength of the signal should not change, only the perceived position of the signal should change.

Therefore, equal-power panning functions are typically used in audio software (and hardware) because perceived signal strength is associated with the combined power of the two channels, and not the combined amplitude.

Functions for −4.5 dB Panning

As a middle option between −6 dB linear panning and −3 dB square-law (or sine-law) panning, some audio software has −4.5 dB panning. This type of panning is a compromise between equal amplitude and equal power. For center panning, the value of the amplitude scalar in the left and right channels is $10^{(-4.5/20)} = 0.5946$. The panning functions for this type are based on modified versions of square-law or sine-law panning.

$$-4.5 \text{ dB modified square-law} = \begin{cases} a_L = (1 - p_N)^{0.75} \\ a_R = (p_N)^{0.75} \end{cases}$$

$$-4.5 \text{ dB modified sine-law} = \begin{cases} a_L = \sqrt{(1 - p_N) \cdot sin\left((1 - p_N) \cdot \frac{\pi}{2}\right)} \\ a_R = \sqrt{p_N \cdot sin\left(p_N \cdot \frac{\pi}{2}\right)} \end{cases}$$

9.2.2 Rhythmic Auto-Pan Effect

A **rhythmic auto-pan effect** is a stereo effect where the amplitude of a signal oscillates between the left and right channels of a stereo signal. It can be created by modulating the amplitude of the left channel opposite of the right channel in a stereo signal. Perceptually, it appears the signal is moving back and forth between the left and right speakers. This effect is essentially the stereo version of tremolo (Section 8.5).

An LFO signal representing values between [−100,100] can be used to perform the automatic, rhythmic panning. The LFO is required to be the same length as the input signal. In this case, each subsequent sample of a signal is panned to a particular position, rather than panning all samples to the same position.

Examples: Create the following function m-file in MATLAB.

```
% AUTOPANEXAMPLE
% This script implements the automatic panning (auto-pan)
% effect. The function "pan" is used to process an
% input signal, along with an array of pan values
% for each sample number in the signal
%
% See also PAN
clc; clear;

% Import test sound file
[in,Fs] = audioread('RhythmGuitar.wav');
N = length(in); Ts = 1/Fs;
% Create time vector t of samples for LFO
t = [0:N-1]*Ts; t = t(:);
f = 1; % Frequency of panning LFO

% Generate sine wave LFO based on f and t
panValue = 100*sin(2*pi*f*t);
panType = 2; % Start with panType = 2, but try 1,2,3,4

% Run pan function
[out] = pan(in,panValue,panType);

sound(out,Fs);
```

Example 9.2: Script: `autoPanExample.m`

9.3 Mid/Side Processing

Mid/side processing is based on the decomposition of a stereo (left and right) signal into separate *mid* and *side* signals. After the signals have been separated, each can be processed independently. During the mastering process of a recorded song, audio engineers may process the instruments or vocals in the center of a mix differently than the instruments on the sides. An example of this would be filtering out low frequencies from the side channels to focus the bass guitar and kick drum in the center of a mix. Another example would be compressing the center of a mix separate from the sides to prevent asymmetrical compression of the stereo field.

9.3.1 Mid/Side Background

When a listener perceives a stereo signal played back over stereo speakers, they do not simply hear the sound coming out of each speaker. A listener can perceive a **phantom center** where the sound appears to be produced between the two speakers. The phantom center occurs when the same signal is presented to both ears simultaneously. If an engineer would like to have a signal be perceived by a listener in the middle, they pan the signal to the center. In this case, an identical mono signal is presented by both the left and right speakers simultaneously.

If an engineer would like to have a signal be perceived on the side, they pan the signal to the left channel or the right channel, and the listener perceives the sound as coming from the same side as the speaker. In this simple case, the original mono signal is only presented in one speaker, but is not presented to the other speaker (in other words: each speaker presents a different signal).

More succinctly, if a signal is perceived in the middle, the left and right channels are identical. If a signal is perceived on the side, the left and right channels are different.

By subtracting the signal on the right channel from the signal on the left channel, any identical signals on both channels cancel out by *destructive interference*. Therefore, any signals panned to the center are removed, leaving only the signal on the *sides*. By adding the signal on the right channel to the signal on the left channel, any identical signals on both channels increase in amplitude by *constructive interference*. Therefore, any signals panned to the center are louder.

9.3.2 Encoding

To encode the mid and side signals from a conventional stereo file with left and right channels, the following equations can be used:

$$x_S[n] = \frac{1}{2} \cdot (x_L[n] - x_R[n])$$

$$x_M[n] = \frac{1}{2} \cdot (x_L[n] + x_R[n])$$

9.3.3 Decoding

To decode, or perfectly recover, the original left and right channels from the mid and side signals, the following equations can be used:

$$x_L[n] = x_M[n] + x_S[n]$$

$$x_R[n] = x_M[n] - x_S[n]$$

Substituting the encoding equations:

$$x_L[n] = x_M[n] + x_S[n] = \left(\frac{1}{2}L + \frac{1}{2}R\right) + \left(\frac{1}{2}L - \frac{1}{2}R\right) = \frac{1}{2}L + \frac{1}{2}L$$

$$x_R[n] = x_M[n] - x_S[n] = \left(\frac{1}{2}L + \frac{1}{2}R\right) - \left(\frac{1}{2}L - \frac{1}{2}R\right) = \frac{1}{2}R + \frac{1}{2}R$$

Therefore, a stereo signal can be decomposed into mid and side signals and still be perfectly reconstructed.

Examples: Create the following function m-file in MATLAB.

```matlab
% MIDSIDEPROCESSING
% This script performs mid/side (sum and difference)
% encoding and decoding.
clc; clear;

[input,Fs] = audioread('stereoDrums.wav');

% Separate stereo signal into two mono signals
left = input(:,1);
right = input(:,2);

% Mid/side encoding
mid = 0.5 * (left + right);
sides = 0.5 * (left - right);

% Add additional processing here
% (e.g., distortion, compression, etc.)
%%%%%%%%%%%%%%%%%%%%%%%%%%%%%%%%%%%%%%%%%%%%%%%%%%%%%%%%%%%%%%%%%%%%




%%%%%%%%%%%%%%%%%%%%%%%%%%%%%%%%%%%%%%%%%%%%%%%%%%%%%%%%%%%%%%%%%%%%

% Mid/side decoding
newL = mid + sides;
newR = mid - sides;

output = [newL , newR];
```

Example 9.3: Script: `midSideProcessing.m`

9.3.4 *Stereo Image Widening*

A stereo signal can be processed using mid/side encoding to change the perceived stereo width. By changing the scaling factor during the encording process, the relative contribution of the mid and side channels can be changed. For scaling factor *w* greater than 1, the stereo image is widened. For scale factor *w* less than 1, the stereo image is narrowed.

$$side = (w) \cdot (left - right)$$

$$mid = (2 - w) \cdot (left + right)$$

Examples: Create the following function m-file in MATLAB.

```
% STEREOIMAGER
% This script demonstrates the stereo image widening effect.
% The effect is based on mid/side processing. The parameter
% "width" can be used to make the example drums file sound
% wider or narrower.
%
% See also MIDSIDEPROCESSING

clc; clear;
[in,Fs] =audioread('stereoDrums.wav');

% Splitting signal into right and left channels
L = in(:,1);
R = in(:,2);

% Create mid and side channels
side = 0.5 * (L-R);
mid = 0.5 * (L+R);

% Width amount (wider if > 1, narrower if < 1)
width = 1.5;

% Scale the mid/side with width
sideNew = width .* side;
midNew = (2 - width) .* mid;

% Create new M/S signal
newLeft = midNew + sideNew;
```

```
newRight = midNew - sideNew;

% Combine signals, concatenated side-by-side, 2 columns
out = [newLeft , newRight];

% Play the sound
sound(out,Fs)
```

Example 9.4: Script: `stereoImager.m`

9.4 Visualizing Stereo Width

The **goniometer** is a visualization used by audio engineers to represent the stereo width of a signal (Hoeg & Grunwald, 1996; Lauridsen & Schlegel, 1956). It is a plot on a polar coordinate system of the combined stereo position of the left and right channels for each time sample, *n*. An example of the goniometer is shown in Figure 9.4.

Values along the vertical axis, at an angle of 90°, represent parts of the signal in the middle (or center) of the stereo field. This occurs when the left and right channels are identical. Conversely, values along the horizontal axis represent parts of the signal when the left and right channels have opposite polarities.

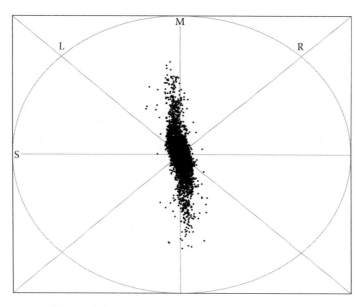

Figure 9.4: Goniometer for stereo drum recording

Values at an angle of 45° represent when there is a signal panned to the right channel while the left channel has zero amplitude. Similarly, values at an angle of 135° represent when there is a signal panned to the left channel while the right channel has zero amplitude.

9.4.1 Polar Coordinates

The stereo position on a goniometer is measured relative to a polar coordinate system. As an alternative to a Cartesian coordinate system that represents position based on orthogonal vertical and horizontal values (rise and run), a polar coordiante system represents position based on a radius (magnitude distance) from the origin and an angle from the horizontal axis. A point on a two-dimensional plane can be represented by either Cartesian coordinates, (x, y), or polar coordinates, (r, θ). In Figure 9.5, the relationship between Cartesian coordinates and polar coordinates is shown.

9.4.2 Goniometer

A goniometer plots a point for each time sample, n, in a stereo signal. The amplitude of the input signal at a time sample for the left channel is $L[n]$ and the right channel is $R[n]$. The position of each point is determined by calculating an angle, θ, between the left and right channels. As a convention, a rotational offset of 45° is included in the angle measurement so the stereo middle has a value of 90°, and the stereo field is symmetric about the vertical axis. Additionally, a combined magnitude, or radius, is calculated as the distance from the origin

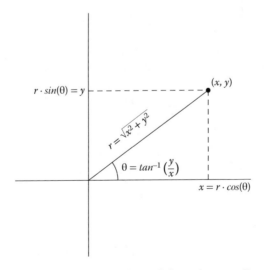

Figure 9.5: Converting polar and Cartesian coordinates

for each point. Finally, the polar coordinates, (r, θ), are converted to Cartesian coordinates, (x, y), for plotting.

$$\theta[n] = tan^{-1}\left(\frac{L[n]}{R[n]}\right) + \frac{\pi}{4}$$

$$r[n] = \sqrt{L[n]^2 + R[n]^2}$$

$$x[n] = r[n] \cdot cos(\theta[n])$$

$$y[n] = r[n] \cdot sin(\theta[n])$$

In MATLAB, inverse tangent, tan^{-1}, should be performed by the function `atan2` in order to have both positive and negative values for the vertical coordinate, y.

In Figure 9.6, several examples are shown for a mono test signal panned to different stereo locations. Additionally, an example is included for when the left and right channels have opposite polarities.

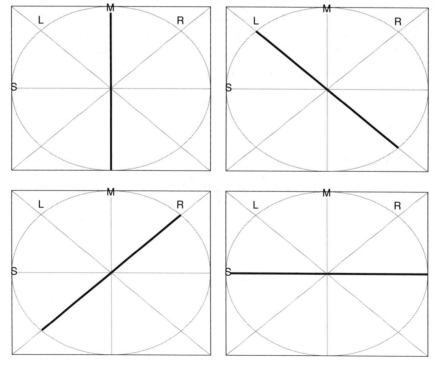

Figure 9.6: Goniometer for a signal panned to different positions

Examples: Create the following m-files in MATLAB.

```
% GONIOMETEREXAMPLE
%
% This script demonstrates several example plots
% of the goniometer.
%
% Examples include a signal panned to the center,
% left, and right. Finally, an example is shown
% for when the left and right channels have opposite
% polarity.
%
% See also GONIOMETER

% Test Signal
Fs = 48000; Ts = 1/Fs;
f = 10; t = [0:Ts:.1].';
x = sin(2*pi*f*t);

% Center
panCenter = [0.707*x , 0.707*x];
subplot(2,2,1);
goniometer(panCenter);

% Left
panLeft = [x , zeros(size(x))];
subplot(2,2,2);
goniometer(panLeft);

% Right
panRight = [zeros(size(x)),x];
subplot(2,2,3);
goniometer(panRight);

% Opposite Polarities
polarity = [0.707 * x , 0.707 * (-x)];
subplot(2,2,4);
goniometer(polarity);
```

Example 9.5: Script: `goniometerExample.m`

```
% GONIOMETER
%
% This function analyzes a stereo audio signal and
% creates a goniometer plot. This visualization indicates
% the stereo width of a signal.
%
% Values along the vertical axis represent parts of
% the signal in the middle (or center) of the
% stereo field. This occurs when the left and right
% channels are identical. Conversely, values along
% the horizontal axis represent parts of the signal
% when the left and right channels have opposite polarities.
%
% Values at an angle of 45 degrees represent when
% there is a signal panned to the right channel and
% the left channel has zero amplitude. Similarly, values
% at an angle of 135 degrees represent when there is
% a signal panned to the left channel and the right
% channel has zero amplitude.
%
% See also GONIOMETEREXAMPLE
function goniometer(in)
N = length(in);
x = size(N,1);
y = size(N,1);

for n = 1:N
    L = in(n,1);
    R = in(n,2);

    radius = sqrt(L^2 + R^2);
    angle = atan2(L,R);
    angle = angle + (pi/4); % Rotate by convention

    x(n,1) = radius * cos(angle);
    y(n,1) = radius * sin(angle);

end
```

```
line([-1 1] ,[-1 1],'Color',[0.75 0.75 0.75]);
line([-1 1] ,[1 -1],'Color',[0.75 0.75 0.75]);
line([0 0] ,[-1 0.95],'Color',[0.75 0.75 0.75]);
line([-0.95 1] ,[0 0],'Color',[0.75 0.75 0.75]);
hold on;
% Circle
th = 0:pi/50:2*pi;
xunit = cos(th);
yunit = sin(th);
plot(xunit, yunit,'Color',[0.75 0.75 0.75]);

% Left
xL = -0.75;
yL = 0.8;
txtL = 'L';
text(xL,yL,txtL,'Color',[0.75 0.75 0.75]);

% Right
xR = 0.73;
yR = 0.8;
txtR = 'R';
text(xR,yR,txtR,'Color',[0.75 0.75 0.75]);

% Mid
xM = -0.018;
yM = 0.96;
txtM = 'M';
text(xM,yM,txtM,'Color',[0.75 0.75 0.75]);

% Mid
xS = -0.98;
yS = 0;
txtS = 'S';
text(xS,yS,txtS,'Color',[0.75 0.75 0.75]);

% Plot Data
plot(x,y,'.b'); axis([-1 1 -1 1]);
hold off;
```

Example 9.6: Function: `goniometer.m`

Bibliography

Hoeg, W., & Grunwald, P. (1996). Visual Monitoring of Multichannel Audio Signals. *Audio Engineering Society Convention*, 100, paper number 4278.

Lauridsen, H., & Schlegel, F. (1956). *Stereofonie und richtungsdiffuse Klangwiedergabe* [Stereophony and directionally diffuse reproduction of sound]. *Gravesaner Blätter, 5*, 28–50.

Chapter 10

Distortion, Saturation, and Clipping
Element-Wise Processing: Nonlinear Effects

10.1 Introduction: Linear and Nonlinear Processing

Methods of processing a signal by multiplying and/or adding a scalar to each element are *linear* types of processing. All processing methods presented previously are based on linear processing. Nonlinear processing is based on methods other than scalar multiplication and addition.

10.1.1 Audio Distortion Effects

Audio distortion effects process signals using a **nonlinear** function. There are many audio effects categorized as distortion effects including overdrive, fuzz, harmonic saturation, and aural exciters. Additionally, many audio technologies process signals in a nonlinear manner such as guitar tube amplifiers, saturated analog tape, and transformers in an audio console. Distortion effects can also be created using DSP. This chapter presents several different algorithms for creating distortion effects digitally.

10.2 Visualizing Nonlinear Processing

Because there are so many types of distortion, it is important for audio engineers to be able to differentiate how each type effects a signal. One way to visualize nonlinear processing is the characteristic curve, or waveshaper curve (Section 6.4.1). For linear types of processing, the slope and *y*-intercept of a straight line are changed on the characteristic curve. As the name suggests, nonlinear functions produce characteristic curves that do not resemble a straight line. Each variation of the characteristic curve represents important differences between distortion effects.

10.2.1 Total Harmonic Distortion Plot

Another method to compare nonlinear processing is the measurment of **total harmonic distortion** (THD). One of the consequences of a distortion effect is the production of signal harmonics. By using a sine wave as the input signal to the distortion effect, the harmonics produced by the nonlinear processing can be analyzed. THD is a comparison between the amplitude of the fundamental frequency of the distorted, sine-wave signal to the amplitude of the other harmonics of the distorted signal (Bates, 1981). Specifically, the standard measurement of THD is based on harmonics 2–6, or the first five harmonics above the fundamental frequency. The following expression can be used to calculate THD based on the amplitude, A_n, for harmonics $n = 1, 2, 3, 4, 5, 6$:

$$THD(\%) = 100 \cdot \sqrt{\frac{(A_2)^2 + (A_3)^2 + (A_4)^2 + (A_5)^2 + (A_6)^2}{(A_1)^2}}$$

A MATLAB function in the Signal Processing Toolbox can be used to calculate and plot THD. The syntax to use the function is the following:

$$r \;=\; \text{thd}(x, Fs, n)$$

- Input Variables
  - x : distorted sine wave
  - Fs : sampling rate
  - n : number of harmonics in calculation
- Output Variable
  - r : total harmonic distortion in dB

Executing the function without an output variable produces a visualization plot of the analysis.

In Example 10.1, the thd function is demonstrated. For this example, assume a hypothetical distortion effect produces an output signal that resembles a square wave when a sine wave is used as the input. As shown in Section 7.3.3, a square wave is a signal with harmonics related to the fundamental frequency. In particular, odd harmonics of the fundamental have a greater amplitude than even harmonics. This result is confirmed by the plot shown in Figure 10.1.

Example: Create the following m-file in MATLAB.

```
% THDEXAMPLE
% This script demonstrates how to analyze and
% visualize total harmonic distortion (THD)
% for a nonlinear function.
%
% See also DISTORTIONEXAMPLE
```

```
clear; clc;
Fs = 48000; Ts = 1/Fs; Nyq = Fs/2;
t = [0:Ts:1-Ts];
f = 2500;
% Input signal
x = sin(2*pi*f*t);
% Hypothetical distortion effect outputs a square wave
y = square(2*pi*f*t);

% Use THD function
thd(y,Fs); % Opens plot
r = thd(y,Fs);  % Returns THD in dB

% Convert dB to percentage
percentTHD = 10^(r/20) * 100
```

Example 10.1: Script: `thdExample.m`

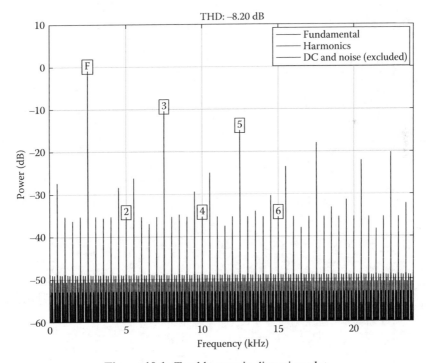

Figure 10.1: Total harmonic distortion plot

Distortion Test Script

The THD measurement and plot, as well as the characteristic curve, are used throughout this chapter to analyze various distortion effects. A script shown in Example 10.2 can be used to test the functions provided for each type of distortion.

Example: Create the following m-file in MATLAB.

```
% DISTORTIONEXAMPLE
% This script is used to test various distortion functions.
% Each algorithm can be analyzed by "uncommenting" the
% code under each section. The waveform, characteristic curve,
% and total harmonic distortion (THD) is plotted for
% each function.
%
% See also THDEXAMPLE

clear;clc;

Fs = 48000; Ts = 1/Fs;
f = 2;         % f = 2 (Waveform and Char Curve). f = 2500 (THD)
t = [0:Ts:1].';
in = sin(2*pi*f*t);  % Used as input signal for each distortion

%%% Infinite clipping
%out = infiniteClip(in);

%%% Half-wave rectification
%out = halfwaveRectification(in);

%%% Full-wave rectification
%out = fullwaveRectification(in);

%%% Hard-clipping
%thresh = 0.5;
%out = hardClip(in,thresh);

%%% Cubic soft-clipping
%a = 1;  % Amount: 0 (no distortion) - 1 (full)
%out = cubicDistortion(in,a);
```

```
%%% Arctangent distortion
%alpha = 5;
%out = arctanDistortion(in,alpha);

%%% Sine distortion
%out = sin((pi/2)*in);

%%% Exponential soft-clipping
%G = 4;
%out = exponential(in,G);

%%% Piece-wise overdrive
%out = piecewise(in);

% Diode clipping
%out = diode(in);

% Asymmetrical distortion
%dc = -0.25;
%out = asymmetrical(in,dc);

%%% Bit crushing
%nBits = 8;
%out = bitReduct(in,nBits);

%%% Dither noise
%dither = 0.003*randn(size(in));
%nBits = 8;
%out = bitReduct(in+dither,nBits);

%%% Plotting
figure(1);   % Use f = 2 (above)
subplot(1,2,1); % Waveform
plot(t,in,t,out); axis([0 1 -1.1 1.1]);
xlabel('Time (sec.)');ylabel('Amplitude');
title('Waveform');

subplot(1,2,2); % Characteristic curve
plot(in,in,in,out); axis([-1 1 -1.1 1.1]);
xlabel('Input Amplitude'); ylabel('Output Amplitude');
```

```
legend('Linear','Distortion'); title('Characteristic Curve');



figure(2); % Total harmonic distortion plot (f = 2500)
thd(out,Fs,5);
axis([0 24 -50 0]);
```

Example 10.2: Script: `distortionExample.m`

10.3 Infinite Clipping

Infinite clipping is a type of distortion effect where the amplitude of a signal is processed to only allow two possible values, the positive and negative full-scale amplitude. All amplitude values greater than zero are processed to equal the positive full-scale amplitude. All amplitude values less than zero are processed to equal the negative full-scale amplitude. In Figure 10.2, the waveform and characteristic curve for infinite clipping are shown. The THD analysis is shown in Figure 10.3.

Example: Create the following function m-file in MATLAB.

```
% INFINITECLIP
% This function implements infinite clipping
% distortion. Amplitude values of the input signal
% that are positive are changed to 1 in the
% output signal. Amplitude values of the input signal
% that are negative are changed to -1 in the
% output signal.
%
% See also HARDCLIP, DISTORTIONEXAMPLE

function [out] = infiniteClip(in)

N = length(in);
out = zeros(N,1);
for n = 1:N
    % Change all amplitude values to +1 or -1 (FS amplitude)
    % "Pin the Rails" (description in audio electronics)
    if in(n,1) >= 0
        % If positive, assign output = 1
        out(n,1) = 1;
    else
```

```
        % If negative, set output = -1
        out(n,1) = -1;


    end


end
```

Example 10.3: Function: `infiniteClip.m`

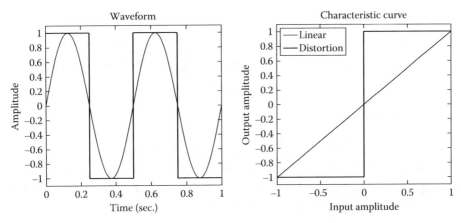

Figure 10.2: Waveform and characteristic curve of infinite clipping

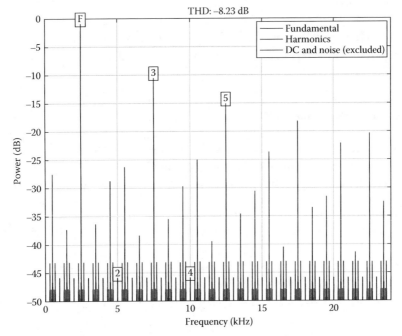

Figure 10.3: THD for infinite clipping with sine wave input, $f = 2500$

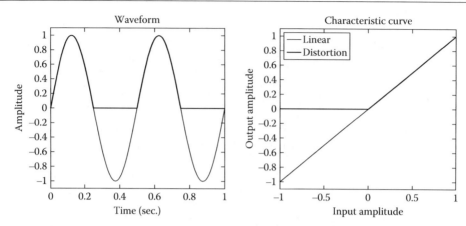

Figure 10.4: Waveform and characteristic curve of half-wave rectification

10.3.1 Sine Wave Analysis

One way to analyze the distortion of the infinite clipping effect is to use a sine wave input signal. By applying infinite clipping to a sine wave, the result is a square wave signal. Therefore, infinite clipping produces harmonic distortion with characteristics resembling the harmonic content of a square wave.

10.4 Rectification

Rectification is a type of distortion effect where the negative amplitude values of a signal are distorted in different ways. Two types of rectification are used in audio: half-wave and full-wave.

10.4.1 Half-Wave Rectification

For **half-wave rectification**, all negative amplitude values are changed to zero. A waveform plot of a signal processed by half-wave rectification and its characteristic curve are shown in Figure 10.4. The THD analysis is shown in Figure 10.5.

Example: Create the following function m-file in MATLAB.

```
% HALFWAVERECTIFICATION
% This function implements full-wave rectification
% distortion. Amplitude values of the input signal
% that are negative are changed to zero in the
% output signal.
```

```
%
% See also FULLWAVERECTIFICATION, DISTORTIONEXAMPLE

function [out] = halfwaveRectification(in)

N = length(in);
out = zeros(N,1);
for n = 1:N

    if in(n,1) ≥ 0
        % If positive, assign input to output
        out(n,1) = in(n,1);
    else
        % If negative, set output to zero
        out(n,1) = 0;

    end

end
```

Example 10.4: Function: halfwaveRectification.m

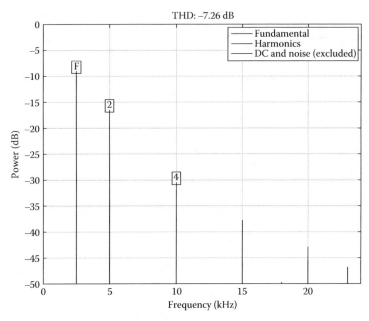

Figure 10.5: THD for half-wave rectification with sine wave input, $f = 2500$

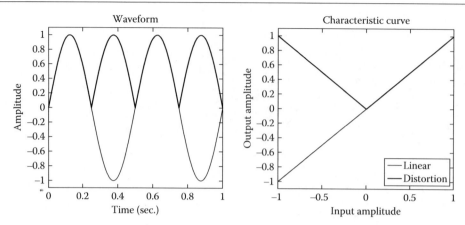

Figure 10.6: Waveform and characteristic curve of full-wave rectification

10.4.2 Full-Wave Rectification

For **full-wave rectification**, the negative amplitude values are changed to the identical positive amplitude values. A waveform plot of a signal processed by full-wave rectification and its characteristic curve are shown in Figure 10.6. The THD analysis is shown in Figure 10.7.

Example: Create the following function m-file in MATLAB.

```
% FULLWAVERECTIFICATION
% This function implements full-wave rectification
% distortion. Amplitude values of the input signal
% that are negative are changed to positive in the
% output signal.
%
% See also HALFWAVERECTIFICATION, DISTORTIONEXAMPLE

function [ out ] = fullwaveRectification(in)

N = length(in);
out = zeros(N,1);
for n = 1:N

    if in(n,1) ≥ 0
        % If positive, assign input to output
        out(n,1) = in(n,1);
    else
        % If negative, flip input
```

```
        out(n,1) = -1*in(n,1);


    end

end
```

Example 10.5: Function: `fullwaveRectification.m`

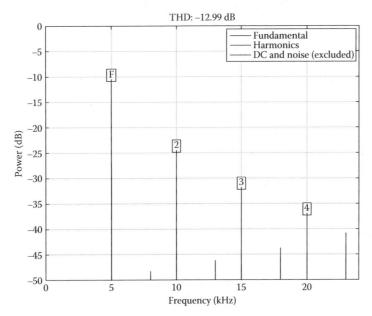

Figure 10.7: THD for full-wave rectification with sine wave input, $f = 2500$

10.5 *Hard Clipping*

Hard clipping is a type of distortion effect where the amplitude of a signal is limited to a maximum amplitude. This effect can be created in analog hardware when a transistor is pushed to a maximum amplitude. At this amplitude, the transistor saturates and cannot output a signal above a specific level, so the input signal is clipped.

This type of distortion effect can also be created using software. It is accomplished by detecting when the amplitude of a signal goes above a specified threshold. If the amplitude of the signal goes above the threshold, then the amplitude is changed to the threshold amplitude (resulting in clipping).

An equivalent process can be performed for negative amplitudes to maintain symmetry. If the amplitude of the signal goes below the negative of the threshold, then that amplitude is changed to the negative threshold amplitude.

An equation to determine the amplitude of a processed or clipped, output signal is the following:

$$y[n] = \begin{cases} thresh & \text{if } x[n] \text{ is } > \text{thresh} \\ x[n] & \text{if } -\text{thresh} < x[n] < \text{thresh} \\ -thresh & \text{if } x[n] \text{ is } < -\text{thresh} \end{cases}$$

Example: Create the following function m-file in MATLAB.

```matlab
% HARDCLIP
% This function implements hard-clipping
% distortion. Amplitude values of the input signal
% that are greater than a threshold are clipped.
%
% Input variables
%   in : signal to be processed
%   thresh : maximum amplitude where clipping occurs
%
% See also INFINITECLIP, PIECEWISE, DISTORTIONEXAMPLE

function [out] = hardClip(in,thresh)

N = length(in);
out = zeros(N,1);
for n = 1:N

    if in(n,1) >= thresh
        % If true, assign output = thresh
        out(n,1) = thresh;
    elseif in(n,1) <= -thresh
        % If true, set output = -thresh
        out(n,1) = -thresh;
    else
        out(n,1) = in(n,1);

    end

end
```

Example 10.6: Function: `hardClip.m`

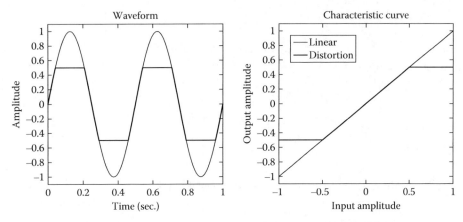

Figure 10.8: Waveform and characteristic curve of hard clipping

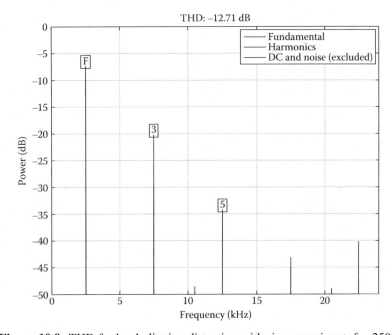

Figure 10.9: THD for hard-clipping distortion with sine wave input, $f = 2500$

Figure 10.8 shows the characteristic curve of a hard-clipping effect with *thresh* = 0.5 FS. The THD analysis is shown in Figure 10.9.

10.6 Soft Clipping

Soft clipping is a type of distortion effect where the amplitude of a signal is saturated along a smooth curve, rather than the abrupt shape of hard clipping. There are many nonlinear functions that can be used to create a soft-clipping distortion effect.

10.6.1 Cubic Distortion

One example of soft-clipping distortion is based on a **cubic function** (Sullivan, 1990). An equation to create soft clipping is:

$$y[n] = x[n] - \frac{1}{3} \cdot (x[n])^3 \qquad (10.1)$$

Example: Create the following function m-file in MATLAB.

```
% CUBICDISTORTION
% This function implements cubic soft-clipping
% distortion. An input parameter "a" is used
% to control the amount of distortion applied
% to the input signal.
%
% Input variables
%    in : input signal
%    a : drive amount (0-1), amplitude of 3rd harmonic
%
% See also ARCTANDISTORTION, DISTORTIONEXAMPLE

function [ out ] = cubicDistortion(in,a)

N = length(in);
out = zeros(N,1);
for n = 1:N

    out(n,1) = in(n,1) - a*(1/3)*in(n,1)^3;

end
```

Example 10.7: Function: `cubicDistortion.m`

By decreasing the scaling coefficient of the cubic term from $\frac{1}{3}$ to 0, the relative amount of distortion is reduced until the function becomes a linear equation. A waveform plot of a sine wave signal that has been processed by a soft-clipping effect and its characteristic curve are shown in Figure 10.10.

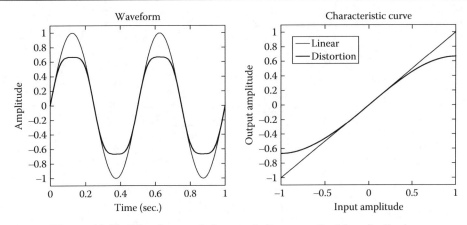

Figure 10.10: Waveform and characteristic curve of cubic soft clipping

Sine Wave Analysis

As was demonstrated in Section 10.3, this distortion effect can be analyzed by using a sine wave input signal, $x[n] = sin(\theta)$. An illustrative result is obtained through the substitution of trigonometric identities.

$$y[n] = x[n] - \frac{1}{3} \cdot (x[n])^3$$

$$y[n] = sin(\theta) - \frac{1}{3} \cdot (sin(\theta))^3$$

$$y[n] = sin(\theta) - \frac{1}{3} \cdot \left(\frac{3 \cdot sin(\theta) - sin(3\theta)}{4} \right)$$

$$y[n] = sin(\theta) - \frac{sin(\theta)}{4} - \frac{sin(3\theta)}{12}$$

$$y[n] = \frac{3}{4} \cdot sin(\theta) - \frac{1}{12} \cdot sin(3\theta)$$

Therefore, the cubic distortion function reduces the amplitude of the input frequency to $\frac{3}{4}$ and produces the third harmonic, 3θ, with an amplitude of $\frac{1}{12}$ of the input signal's amplitude. This result is confirmed by the THD analysis shown in Figure 10.11.

10.6.2 Arctangent Distortion

A different method for creating soft-clipping distortion is based on the **arctangent function** (Smith, 2010).

$$y[n] = \frac{2}{\pi} \cdot arctan(\alpha \cdot x[n])$$

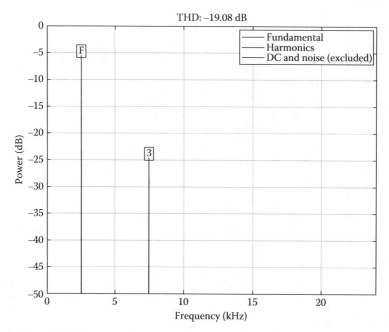

Figure 10.11: THD for soft-clipping, cubic distortion with sine wave input, $f = 2500$

In Example 10.8, a function for create arctangent distortion is demonstrated. A plot of a sine wave processed by the function and the distortion's characteristic curve are shown in Figure 10.12.

Example: Create the following function m-file in MATLAB.

```
% ARCTANDISTORTION
% This function implements arctangent soft-clipping
% distortion. An input parameter "alpha" is used
% to control the amount of distortion applied
% to the input signal.
%
% Input variables
%    in : input signal
%    alpha : drive amount (1-10)
%
% See also CUBICDISTORTION, DISTORTIONEXAMPLE

function [out] = arctanDistortion(in,alpha)

N = length(in);
out = zeros(N,1);
```

```
for n = 1:N

    out(n,1) = (2/pi)*atan(in(n,1)*alpha);

end
```

Example 10.8: Function: `arctanDistortion.m`

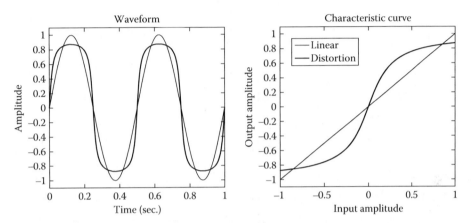

Figure 10.12: Waveform and characteristic curve of arctangent distortion

The relative amount of distortion can be set by the α coefficient. Typical values of the α coefficient are in a range of $\alpha \in [1, 10]$. For values of $\alpha \gg 10$, the soft-clipping distortion approaches infinite clipping distortion. The characteristic curve of arctangent distortion is shown in Figure 10.13 for various values of α. In Figure 10.14, the THD of the arctangent distortion is shown for $\alpha = 5$.

Power Series Approximation of the Arctangent Function

As shown in Section 10.6.1 with the use of trigonometric identities, soft-clipping distortion produces odd harmonics with decreasing amplitude as frequency increases. As another specific demonstration for arctangent distortion, the function can be approximated by a **power series** of odd positive integers. In this power series, the value of the input signal, $x[n]$, is raised to the power of odd whole numbers. The arctangent function is approximated by a linear combination of these terms.

$$arctan(x[n]) = x[n] - \frac{(x[n])^3}{3} + \frac{(x[n])^5}{5} - \frac{(x[n])^7}{7} + \frac{(x[n])^9}{9} \cdots$$

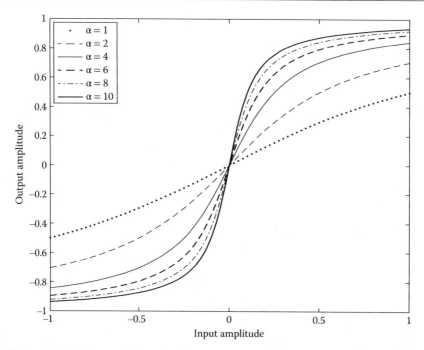

Figure 10.13: Characteristic curve for arctangent distortion with various α values

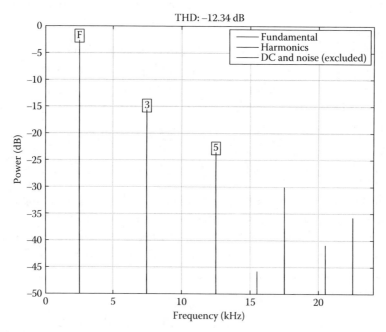

Figure 10.14: THD for arctangent distortion with sine wave input, $f = 2500$

10.6.3 Additional Clipping Functions

Several other mathematical functions can be used to create variations of clipping distortion. Each function shares a similar characteristic curve with a nonlinear function presented previously and represents a modified approach for saturation effects.

Sine Distortion

The sine function has been shown to be used for signal synthesis (Section 7.3.1). It can also be used as an audio effect to create distortion. In this case, the amplitude values of an input signal are processed by the sine function, $y[n] = sin(2 \cdot \pi \cdot x[n])$. The result can be seen in the characteristic curve in Figure 10.15. The values of the input signal are mapped along a smooth soft-clipping curve of a single cycle of the sine function. A plot of the THD for sine distortion is shown in Figure 10.16.

Exponential Soft Clipping

An exponential function can be used to create another variation of the soft-clipping effect. A gain parameter, $G \in [1, 10]$, can be used to change the amount of drive for the effect.

$$y[n] = \begin{cases} (1 - e^{-|G \cdot x[n]|}) & x[n] \geq 0 \\ -(1 - e^{-|G \cdot x[n]|}) & x[n] < 0 \end{cases}$$

An additional term, $\frac{x[n]}{|x[n]|}$, can be included in the expression to multiply by 1 when $x[n] \geq 0$ or multiply by -1 when $x[n] < 0$. Therefore, a conditional statement would not be necessary and a single expression can be used for the entire operating range.

$$y[n] = \frac{x[n]}{|x[n]|} \cdot (1 - e^{-|G \cdot x[n]|})$$

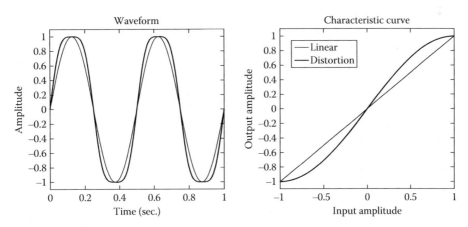

Figure 10.15: Waveform and characteristic curve of sine distortion

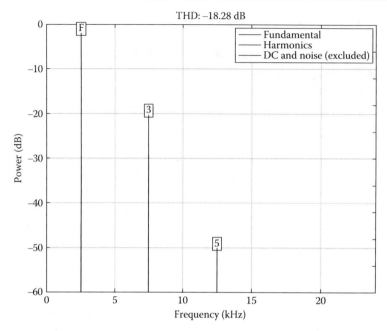

Figure 10.16: THD for sine distortion with sine wave input, $f = 2500$

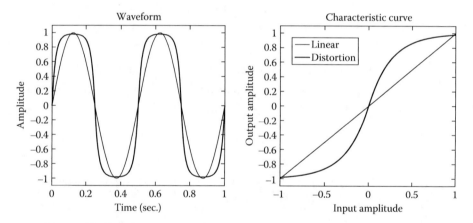

Figure 10.17: Waveform and characteristic curve of exponential distortion

In Example 10.9, a function to perform exponential distortion is demonstrated. The waveform and characteristic curve of exponential distortion effect are shown in Figure 10.17, along with the THD analysis in Figure 10.18.

Examples: Create the following function m-file in MATLAB.

```
% EXPONENTIAL
% This function implements exponential soft-clipping
% distortion. An input parameter "G" is used
```

```
% to control the amount of distortion applied
% to the input signal.
%
% Input variables
%   in : input signal
%   G : drive amount (1-10)
%
% See also CUBICDISTORTION, ARCTANDISTORTION, DISTORTIONEXAMPLE

function [ out ] = exponential(in,G)

N = length(in);
out = zeros(N,1);
for n = 1:N

    out(n,1) = sign(in(n,1)) * (1 - exp(-abs(G*in(n,1))));

end
```

Example 10.9: Function: `exponential.m`

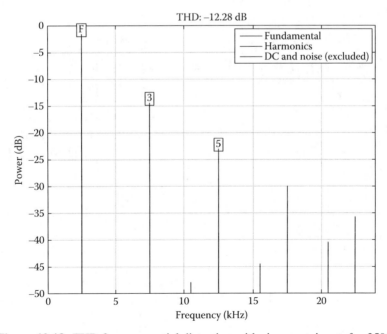

Figure 10.18: THD for exponential distortion with sine wave input, $f = 2500$

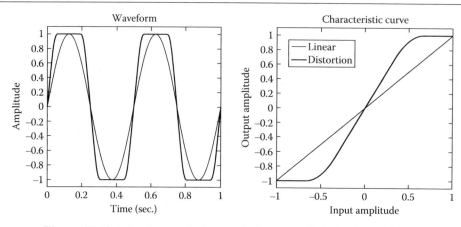

Figure 10.19: Waveform and characteristic curve of piece-wise overdrive

Piece-Wise Overdrive

To simulate how many audio systems are linear within an operating region and nonlinear outside of that region, a piece-wise function can be used (Dutilleux et al., 2002). While the amplitude of the input signal is near zero, the piece-wise function is linear. As the input signal increases in amplitude, it reaches a nonlinear region of the piece-wise function. This introduces soft-clipping distortion. The final region of the piece-wise function while the input signal is near full-scale amplitude adds hard clipping. This ensures the input signal is bounded within the full-scale range.

$$
y[n] = \begin{cases} 2 \cdot x[n] & 0 < |x[n]| < \frac{1}{3} \\ \frac{x[n]}{|x[n]|} \cdot \frac{3-(2-3 \cdot |x[n]|)^2}{3} & \frac{1}{3} < |x[n]| < \frac{2}{3} \\ \frac{x[n]}{|x[n]|} & \frac{2}{3} < |x[n]| \end{cases}
$$

In Example 10.10, a function to perform piece-wise overdrive is demonstrated. The waveform and characteristic curve of the effect are shown in Figure 10.19, along with the THD analysis in Figure 10.20.

Examples: Create the following function m-file in MATLAB.

```
% PIECEWISE
% This function implements a piecewise distortion
% algorithm. Within one operating region, the
% input signal is not distorted. When the signal
% is outside of that operating region, it is clipped.
%
% See also HARDCLIP, DISTORTIONEXAMPLE

function [ out ] = piecewise(in)
```

```
out = zeros(size(in));

for n = 1:length(in)
    if abs(in(n,1)) ≤ 1/3
        out(n,1) = 2*in(n,1);
    elseif abs(in(n,1)) > 2/3
        out(n,1) = sign(in(n,1));
    else
        out(n,1) = sign(in(n,1))*(3-(2-3*abs(in(n,1)))^2)/3;
    end
end
```

Example 10.10: Function: `piecewise.m`

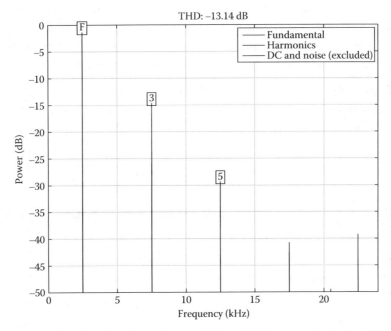

Figure 10.20: THD for piecewise overdrive with sine wave input, $f = 2500$

Diode Clipping

A circuit component found in audio electronics is the **diode**. It can be used for nonlinear processing and is found in many distortion guitar pedals and amplifiers. The **Shockley diode equation** (Shockley, 1949) is a mathematical formula for describing the processing of a diode: $I = I_s \cdot \left(e^{\frac{V_D}{\eta \cdot V_T}} - 1 \right)$.

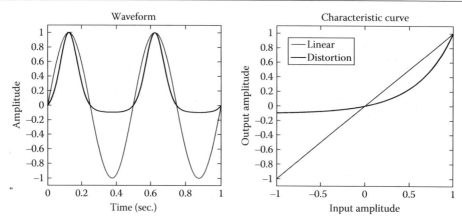

Figure 10.21: Waveform and characteristic curve of diode clipping

The input signal is the voltage across the diode, V_D. The output signal is the current through the diode, I. The remaining variables can be given constant values. The thermal voltage, V_T, is 25.3 mV. The emission coefficent, η, is between 1 and 2 for different types of diodes. The saturation current, I_s, is very small and represents negative bias leakage current.

A modified version of the Shockley diode equation can be used for processing digital signals within a full-scale range. By selecting an emission coefficent, $\eta = 1.68$, and a saturation current, $I_s = .105$, the nonlinear processing of Figure 10.21 can be accomplished.

$$y[n] = I_s \cdot \left(e^{\frac{0.1 \cdot x[n]}{\eta \cdot V_T}} - 1 \right)$$

In Example 10.11, a function to perform diode clipping is demonstrated. The waveform and characteristic curve of the effect are shown in Figure 10.21, along with the THD analysis in Figure 10.22.

Examples: Create the following function m-file in MATLAB.

```
% DIODE
% This function implements the Shockley
% ideal diode equation for audio signals
% with an amplitude between -1 to 1 FS
%
% See also ASYMMETRICAL, DISTORTIONEXAMPLE

function [out] = diode(in)
```

```
% Diode characteristics
Vt = 0.0253; % thermal voltage
eta = 1.68; % emission coefficient
Is = .105;  % saturation current


N = length(in);
out = zeros(N,1);


for n = 1:N


    out(n,1) = Is * (exp(0.1*in(n,1)/(eta*Vt)) - 1);


end
```

Example 10.11: Function: `diode.m`

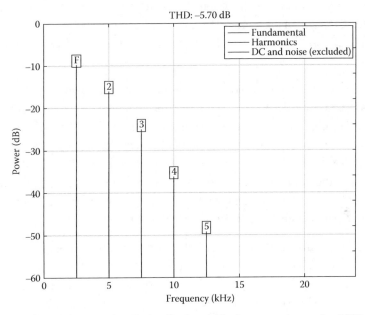

Figure 10.22: THD for diode clipping with sine wave input, $f = 2500$

10.7 Bit reduction

Bit reduction is a type of distortion effect where the number of possible amplitude values of a signal is reduced. This distortion occurs when an audio signal with a higher bit depth is converted to a lower bit depth, like when a musical recording is mastered for CD quality. If done correctly, this bit reduction process should be inaudible. This type of distortion can also be used as an extreme, audible effect called **bit crushing**. In this case, the number of possible amplitude values is greatly reduced, much lower than CD quality.

10.7.1 Bit Depth Background

At a fundamental level, information in a computer is represented as a collection of 1s and 0s. This way of representing information is called a **binary number system**. A digit in a binary number system is called a bit, which is a placeholder for either a 1 or 0 value. As a comparison, a digit in a decimal number system can have one of ten possible values, 0–9.

In a binary number system, a single bit can only represent two possible values. By combining together multiple bits to represent a single value, the are more possibilities. By using two bits together, there are four possible combinations of 1s and 0s. Therefore, four values can be represented by two bits. The relationship between possible values, v, and the number of bits, η, is $v = 2^{\eta}$.

When an audio signal is recorded, its amplitude is measured, or sampled. Each sample is a numerical value to be stored in a computer. At a single point in time, an audio signal could have an amplitude with a large value or an amplitude with a small value. In other words, it is necessary for a computer to be able to represent a wide range of possible amplitude values for each sample. However, because a computer uses a finite number of bits to represent information, it is not possible to have an infinite range, or infinite precision, for an amplitude value.

The process of measuring the amplitude of an audio signal and setting it to one of the possible digital values is called **quantization**. Due to the finite range and precision of digital information, analog amplitude is rounded to a discrete value. In effect, the rounding process adds low-level noise to the sampled signal called quantization noise, or quantization distortion. Therefore, the quantization process sets a digital noise floor for a sampled signal, somewhat similar to the inherent noise floor in many analog systems. The noise floor is the minimum amplitude level that can be represented. Any signal lower in amplitude is masked by the noise floor (i.e., rounded to an amplitude value of zero).

The relative level of the digital noise floor is determined by the number of bits used to represent an amplitude value. A common bit depth for digital audio is $\eta = 16$ bits (Section 2.5.1). Each sample is represented by a combination of sixteen 1s and 0s. For this bit depth, there are $v = 2^{16} = 65,536$ possible amplitude values for each sample. The dynamic range on the decibel scale between the maximum and minimum possible amplitude value is $20 \cdot log_{10}(\frac{65536}{1}) = 96$ dB. This is typically a sufficient dynamic range for digital signals because it exceeds the dynamic range of most analog audio signals and playback systems.

In situations where a greater dynamic range is desired, another common bit depth for digital audio can be utilized: $\eta = 24$ bits. For this bit depth, the dynamic range on the decibel scale is $20 \cdot log_{10}(\frac{2^{24}}{1}) = 144$ dB, which exceeds the dynamic range of human hearing.

Although a computer represents information using a binary number system, a MATLAB programmer does not need to deal with the complexity of it. Instead, binary numbers are surreptitiously converted and displayed as intuitive, decimal values. Nonetheless, the use of binary numbers provides the conceptual foundation for performing bit reduction as an audio effect.

10.7.2 Bit-Reduction Algorithm

Nonlinear processing can be used to reduce the number of possible amplitude values for a signal. Consider an extreme case of infinite clipping (Section 10.3), where the processed signal only has two possible values: -1 or 1. This type of nonlinear processing has the same number of possible amplitude values as 1-bit audio.

For bit-reduction distortion, an audio engineer chooses the number of bits, η, then the processed signal is restricted to 2^η possible amplitude values. Bit reduction can be implemented as an extreme effect, similar to infinite clipping, when only a low number of bits are used. It can also be implemented to simulate primative digital systems, which were only capable of 8-, 10-, or 12-bit audio. Examples include processing synthesized signals to simulate early video game audio and processing the output of a reverberation effect to simulate early digital reverb units.

The process of restricting the number of possible amplitude levels can be accomplished by using rounding functions (e.g., `round`, `fix`, `floor`, `ceil`). A limitation of the rounding functions in MATLAB is that they only round to integer numbers. If an audio signal is within the full-scale range of -1 to 1, these functions only round a signal's amplitude to three possible integers (i.e., -1, 0, and 1). Therefore, additional processing is necessary to have more possible amplitude values.

An approach to perform bit reduction starts by temporarily increasing the amplitude of a signal, so it is stretched across an appropriate number of integer values. This may involve stretching the signal across 4, 8, 16, ..., 65,536, etc. integers. Then, one of the rounding functions is used to remove any fractional values. Finally, the amplitude of the processed signal is decreased to shrink the final signal within the necessary full-scale range.

Examples: Create the following function m-file in MATLAB.

```
% BITREDUCT
% This function creates a bit reduction or
% bit crushing distortion. It uses an input
% variable, "nBits", to determine the number of
% amplitude values in the output signal. This
% algorithm can have a fractional number of bits.
```

```
% similar to the processing found in some audio plug-ins.
%
% Input variables
%   inputSignal : array of samples for the input signal
%   nBits : scalar for the number of desired bits
%
% See also DISTORTIONEXAMPLE, ROUND, CEIL, FLOOR, FIX

function [out] = bitReduct(in,nBits)

% Determine the desired number of possible amplitude values
ampValues = 2 ^ nBits;

% Shrink the full-scale signal (-1 to 1, peak-to-peak)
% to fit within a range of 0 to 1
prepInput = 0.5*in + 0.5;

% Scale the signal to fit within the range of the possible values
scaleInput = ampValues * prepInput;

% Round the signal to the nearest integers
roundInput = round(scaleInput);

% Invert the scaling to fit the original range
prepOut = roundInput / ampValues;

% Fit in full-scale range
out = 2*prepOut - 1;
```

Example 10.12: Function: `bitReduct.m`

A plot of a sine-wave signal reduced to 2 bits and the characteristic curve of the process is shown in Figure 10.23.

Similarly, a plot of a sine-wave signal reduced to 4 bits and the characteristic curve of the process is shown in Figure 10.24.

As with other types of distortion, there are harmonics produced as a result of performing bit reduction. The relative amplitude of the harmonics depends on the number of bits used in the process. The amplitude of the harmonics is greater when fewer bits are used. In Figure 10.25, the total harmonic distortion is shown for a sine-wave signal reduced to 8 bits.

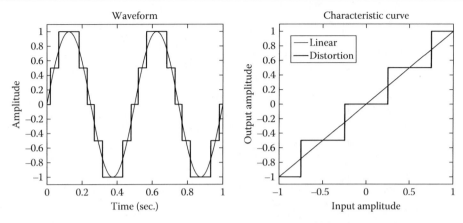

Figure 10.23: Waveform and characteristic curve of bit reduction: 2 bits

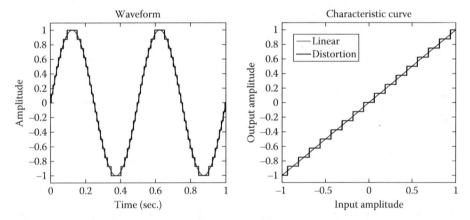

Figure 10.24: Waveform and characteristic curve of bit reduction: 4 bits

10.7.3 Dither Noise

When a musical recording is mastered and finalized, one part of the process is to perform bit reduction to CD quality, 16 bits. Ideally, this task would be inaudible for a listener. Nonetheless, bit reduction creates distortion, or quantization noise, in the resulting signal. The harmonics produced by bit reduction are at frequencies that are multiples of the original signal. Another way to describe this relationship is the quantization noise is correlated with (i.e., related to) the original signal.

Coincidentally, listeners are perceptually adept at detecting harmonically related parts of complex signals. This leads to a greater possibility the bit reduction process produces audible distortion. To avoid this issue, low-level **dither noise** is added to the signal prior to bit reduction (Vanderkooy & Lipshitz, 1987). Because of this, the quantization distortion is not harmonically related to the original signal. In other words, the resulting noise is intentionally uncorrelated, making it harder for a listener to perceive.

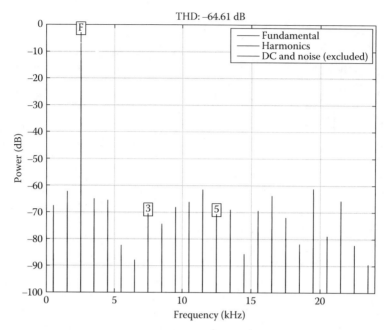

Figure 10.25: THD for bit reduction with sine wave input, $f = 2500$

Dither noise can either be white noise (Section 7.4.1) or spectrally shaped noise to make it even less likely for a listener to perceive. The amplitude of the noise is typically set to the amplitude of 1 bit or $\frac{1}{2}$ bit.

To visualize the result of dithering, the THD spectrum for a sine wave with white noise is shown in Figure 10.26 for bit reduction to 8 bits. Compared to Figure 10.25 without dither, the additional noise spreads out the quantization distortion across the spectrum. Instead of having a few harmonics with a relatively high amplitude, the dithered signal has a lower amplitude at harmonically related frequencies.

10.8 Harmonic Analysis of Distortion Effects

There is an important relationship between the resulting harmonics of a distortion effect and the nonlinear function used in processing. Some distortion effects cause odd harmonics to have a larger amplitude. Examples include infinite clipping, hard clipping, cubic distortion, arctangent distortion, exponential distortion, and bit crushing. Other distortion effects cause even harmonics to have a larger amplitude. One example is full-wave rectification. There are also other types of distortion that produce both odd and even harmonics, like diode clipping and half-wave rectification. In general, an audio engineer can determine the harmonic structure of a distortion effect by considering whether the nonlinear algorithm is an *even* or *odd* mathematical function.

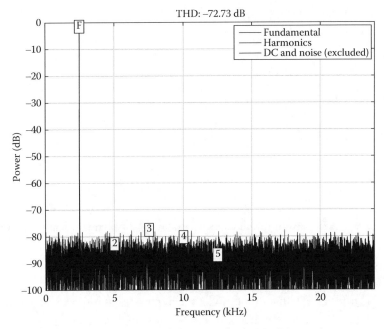

Figure 10.26: THD for bit reduction with dither noise

10.8.1 Even and Odd Mathematical Functions

One category of mathematical functions are called **even functions**. These functions are defined by the following relationship: $f(-x) = f(x)$. In other words, the function produces an identical result for an input value whether it is positive or negative. As an example: $|-5| = |5|$. Therefore, the absolute value function is an example of an even function. Another way to define an even function is to consider its characteristic curve. As shown in Figure 10.6, the absolute value function is symmetric, or a reflected mirror image, about the y-axis. For a Cartesian coordinate system, quadrant I is symmetric with quadrant II and quadrant III is symmetric with quadrant IV. This symmetry is true for all even functions. A nonlinear function that meets the requirements of an even function produces even order harmonics.

Another category of mathematical functions is called **odd functions**. These functions are defined by the following relationship: $f(-x) = -f(x)$. In other words, when a negative input value is used with the function it produces an identical result as when a positive input value is used and the result is multiplied by -1. As an example: $(-x)^3 = -(x^3)$. Similar to even functions, another way to define an odd function is to consider its characteristic curve. As shown in Figure 10.8, the hard-clipping distortion function has a rotational, or circular, symmetry about the origin. For a Cartesian coordinate system, quadrant I is symmetric with quadrant III and quadrant II is symmetric with quadrant IV. This symmetry is true for all odd functions. A nonlinear function that meets the requirements of an odd function produces odd order harmonics.

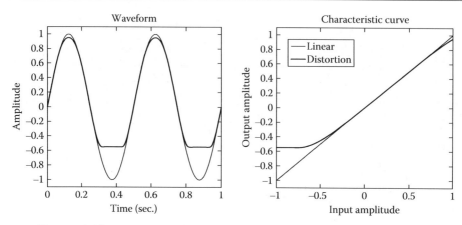

Figure 10.27: Waveform and characteristic curve of asymmetrical distortion

10.8.2 Asymmetrical Distortion Functions

There are nonlinear functions that do not meet the symmetrical requirements of even or odd functions. Two examples include half-wave rectification and diode clipping. The characteristic curves for these functions (Figures 10.4 and 10.21) are not symmetric about the *y*-axis or the origin. When used as distortion effects, these functions produce a combination of even and odd harmonics.

Including a DC Offset with a Distortion Function

One approach to modifying a nonlinear function is to introduce a DC offset prior to distortion and then remove it afterwards. This creates an asymmetrical characteristic curve, and leads to both even and odd harmonics at the output. In Example 10.13, a function to perform asymmetrical distortion is demonstrated. The waveform and characteristic curve of the effect are shown in Figure 10.27, along with the THD analysis in Figure 10.28.

Example: Create the following function m-file in MATLAB.

```
% ASYMMETRICAL
% This function creates a distortion effect
% that is neither "even" or "odd." Therefore,
% the resulting signal has both even and
% odd harmonics.
%
% Input variables
%   in : input signal
%   dc : offset amount
%
```

```
% See also CUBICDISTORTION, DISTORTIONEXAMPLE

function out = asymmetrical(in,dc)
N = length(in);
x = in + dc; % Introduce DC offset
y = zeros(N,1);

for n = 1:N
    % Conditional to ensure "out" is a
    % monotonically increasing function
    if abs(x(n,1)) > 1
        x(n,1) = sign(x(n,1));
    end

    % Nonlinear distortion function
    y(n,1) = x(n,1) - (1/5)*x(n,1)^5;

end
out = y - dc; % Remove DC offset
```

Example 10.13: Function: `asymmetrical.m`

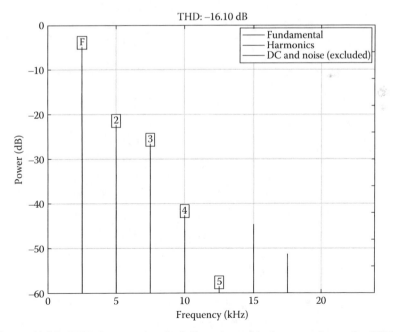

Figure 10.28: THD for asymmetrical distortion with sine wave input, $f = 2500$

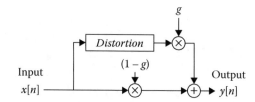

Figure 10.29: Block diagram of parallel distortion

10.9 *Parallel Distortion*

A common feature of some software distortion effects is the ability to blend between the *dry*, unprocessed signal and the *wet*, distorted signal. This feature can be created by using parallel distortion. Typically a variable parameter is used to blend from 0%–100% wet. The blend percentage can be converted to a linear amplitude scalar, *g*, by dividing the percentage by 100. An example of the block diagram for this effect is shown in Figure 10.29.

Any of the various types of distortion can be used on the wet path of the parallel distortion. In Example 10.14, arctangent soft clipping is used as the type of distortion.

Examples: Create and run the following m-file in MATLAB.

```
% PARALLELDISTORTION
% This script demonstrates how to create parallel
% distortion. It allows for the "dry" unprocessed
% signal to be blended with the "wet" processed
% signal.
%
% See also ARCTANDISTORTION

clear; clc;close all;

[in,Fs] = audioread('AcGtr.wav');

% Alpha - amount of distortion
alpha = 8;

% Wet path - distortion
dist = arctanDistortion(in,alpha);

% Pick an arbitrary "wet/dry mix" value
mix = 50;   % Experiment with values from 0 - 100
```

```
% Convert to a linear gain value
g = mix/100;

% Add together "wet/dry" signals
out = g * dist + (1-g) * in;

% Listen to Result
sound(out,Fs);
```

Example 10.14: Script: parallelDistortion.m

Bibliography

Bates, C. M. (1981). Total harmonic distortion measurements in magnetic tape equipment. *Audio Engineering Society Convention*, 70.

Dutilleux, P., & Zölzer, U. (2002). Nonlinear processing. In U. Zölzer (Ed.), *DAFX: Digital audio effects* (p. 118). New York, NY: John Wiley & Sons.

Shockley, W. (1949). The theory of p-n junctions in semiconductors and p-n junction transistors. *The Bell System Technical Journal, 28*(3), 435–489.

Smith, J. O. (2007) *Physical audio signal processing*, W3K Publishing, www.books.w3k.org/, ISBN 978-0-9745607-2-4.

Sullivan, C. R. (1990). Extending the Karplus-Strong algorithm to synthesize electric guitar timbres with distortion and feedback. *Computer Music Journal, 14*, 26–37.

Vanderkooy, J., & Lipshitz, S. P. (1987). Dither in digital audio. *Journal of the Audio Engineering Society, 35*(12), 966–975.

Chapter 11

Echo Effects
Systems with Memory: Audible Delay

11.1 Introduction: Systems with Memory

The systems described from Chapters 6 to 10 are all based on element-wise processing. An equivalent categorization is to describe the systems as *memoryless systems*. In other words, the output at any time from the system only depends on the input to the system at the same time. These systems do not require the memory of previous time samples to process a signal.

There are many other signal processing systems used in audio where the output at any time can depend on the input signal at a previous time. These systems require memory to store the input samples for later use. These are *systems with memory*.

11.2 Delay

Systems with memory have the ability to temporarily store a signal and delay it over time. A **delay** processing block receives a sample of the input signal at sample number n and stores this sample in memory for some number of samples, d. At sample number $n + d$, the output of a delay processing block is the input sample, $x[n]$. Therefore, when input sample $x[n]$ is received by a delay processing block, the output of the block at the same time is the sample $x[n - d]$, or the sample stored in memory for d samples. A block diagram for a basic system with delay is shown in Figure 11.1.

11.2.1 Series Delay

If two or more delay blocks are placed in series, the final output can be written as a combination of the individual effects. In other words, a single equivalent delay block can be substituted in place of multiple series delay blocks to achieve identical processing. A visual description is provided in Figure 11.2.

Input
x[n]
d-Samples
Delay
Ouput
y[n] = x[n − d]

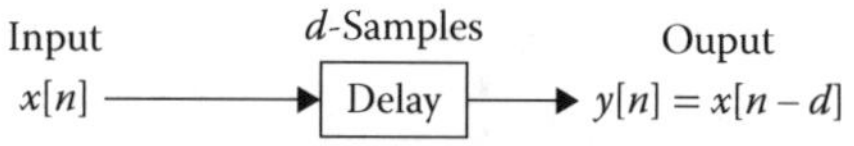

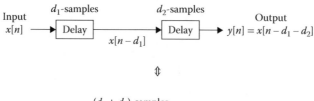

Figure 11.1: Block diagram of delay

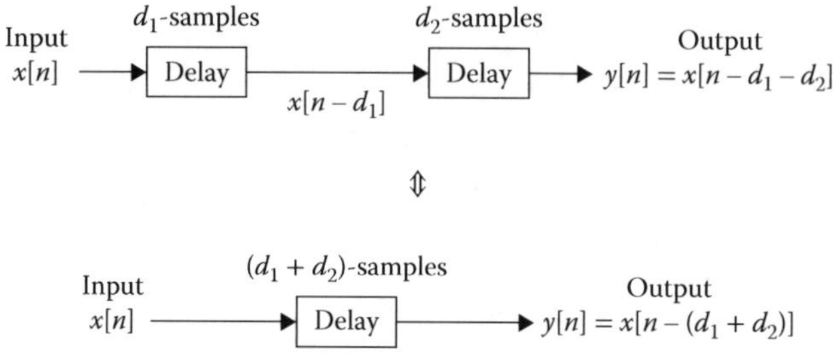

Input
x[n]
d1-samples
Delay
x[n − d1]
d2-samples
Delay
Output
y[n] = x[n − d1 − d2]
⇕
Input
x[n]
(d1 + d2)-samples
Delay
Output
y[n] = x[n − (d1 + d2)]

Figure 11.2: Block diagram of series delay

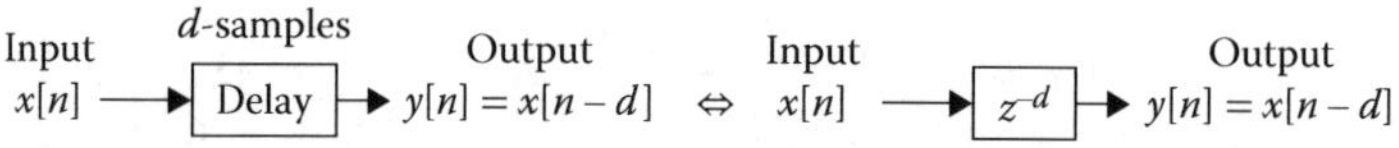

Input
x[n]
d-samples
Delay
Output
y[n] = x[n − d]
⇔
Input
x[n]
z-d
Output
y[n] = x[n − d]

Figure 11.3: Block diagram delay notation

11.2.2 Block Diagram Delay Notation

In order to simplify the display of a block diagram, the notation z^{-d} is used to represent a delay block of d samples (Figure 11.3).

11.3 Converting Delay Time to Samples

When creating delay effects it is often necessary to perform a calculation to convert between different time units. When indexing an array of samples, it is necessary to use a unit of samples. In some situations, audio engineers also consider time in seconds (or milliseconds) and relative to the syncronized tempo of a song (beats per minute).

11.3.1 Converting Seconds to Samples

The relationship between a time in seconds and a time in samples is contained in the sampling rate for a signal. The sampling rate can be used to convert samples to seconds and vice versa. Example 11.1 demonstrates the conversion from seconds to samples and milliseconds to samples.

Examples: Create and run the following m-file in MATLAB.

```
% CONVERTSECSAMPLES
% This script provides two examples for converting a time delay
% in units of seconds to samples and milliseconds to samples
%
% See also CONVERTTEMPOSAMPLES


% Example 1 - Seconds to Samples


Fs = 48000; % arbitrary sampling rate

timeSec = 1.5; % arbitrary time in units of seconds

% Convert to units of samples
timeSamples = fix(timeSec * Fs); % round to nearest integer sample


% Example 2 - Milliseconds to Samples


timeMS = 330; % arbitrary time in units of milliseconds

% Convert to units of seconds
timeSec = timeMS/1000;

% Convert to units of samples
timeSamples = fix(timeSec * Fs); % round to nearest integer sample
```

Example 11.1: Script: `convertSecSamples.m`

11.3.2 Converting Tempo to Samples

One technique used by audio engineers with echo effects is to synchronize the delay time with the tempo of a song. Therefore, the signal repetitions occur in rhythm with a performance. The synchronized delay time is calculated by converting tempo, in beats per minute, to samples using the following relationships.

$$X\frac{beats}{minute} \cdot \frac{1\ minute}{60\ sec} = \frac{beats}{sec}$$

$$\frac{1}{\frac{beats}{sec}} = \frac{sec}{beat}$$

$$\frac{sec}{beat} \cdot \frac{F_s\ samples}{sec} = Y\frac{samples}{beat}$$

Examples: Create and run the following m-file in MATLAB.

```
% CONVERTTEMPOSAMPLES
% This script provides an example for calculating a delay time
% in units of samples that will be synchronized with the tempo
% of a song in units of beats per minutes (BPM).
%
% Assume a (4/4) time signature where a BEAT = QUARTER NOTE
%
% See also CONVERTSECSAMPLES

% Example - Convert Tempo Sync'd Delay to Samples

Fs = 48000; % arbitrary sampling rate

beatsPerMin = 90; % arbitrary tempo in units of beats per minute

% Calculate beats per second
beatsPerSec = beatsPerMin / 60; % 1 minute / 60 seconds

% Calculate # of seconds per beat
secPerBeat = 1/beatsPerSec;

% Note division
% 4 = whole, 2 = half, 1 = quarter, 0.5 = 8th, 0.25 = 16th
noteDiv = 1 ;

% Calculate delay time in seconds
timeSec = noteDiv * secPerBeat;
```

```
% Convert to units of samples
timeSamples = fix(timeSec * Fs); % round to nearest integer sample
```

Example 11.2: Script: `convertTempoSamples.m`

11.4 Categorizing Echo Effects

This chapter discusses systems with delay that create echo effects. These effects are categorized by the creation of an audible delay perceived by listeners as an echo.

11.4.1 Perceptual Temporal Fusion

Temporal fusion is a perceptual phenomenon for listeners. The basic premise of the phenomenon is listeners perceptually fuse signal repetitions with a short time delay into a single auditory image.

An important factor in temporal fusion is a listener's **echo threshold**. When the time delay between repetitions is less than the echo threshold, one sound is perceived instead of distinct, delayed repetitions. When the time delay between repetitions is longer than the echo threshold, multiple echoes are perceived.

The time delay of a listener's echo threshold is ~30 ms. The precise time delay depends on several things including the attack, decay, sustain, release (ADSR) envelope of the signal, the relative amplitude of the repetitions, and the perceived stereo separation of the repetitions. This chapter focuses on systems with a time delay longer than 30 ms, while Chapter 12 discusses systems with a time delay shorter than a listener's echo threshold.

11.5 Feedforward Echo

An echo effect is created by using parallel processing such that a delayed signal is combined with the original version. Audio engineers refer to the processed delayed signal as the *wet signal* and the parallel delayed path as the *wet path*, or a **delay line**. Similarly, the unprocessed signal is called the *dry signal* on the *dry path*.

One basic block diagram for a system with time delay is shown in Figure 11.4. This system is described as a feedforward echo because the input is fed to the output on a forward, delayed path.

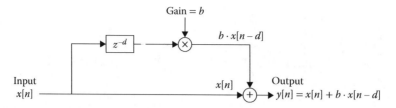

Figure 11.4: Block diagram of feedforward echo

11.5.1 Difference Equation

The output of a system, $y[n]$, can be written as a function of the input signal, $x[n]$, and past samples from delay blocks. This way of representing the system is called the **difference equation**. Describing a block diagram by a difference equation provides a path toward implementing the system in computer code.

In Example 11.3, consider the system in Figure 11.4 with difference equation $y[n] = x[n] + b \cdot x[n-d]$. A feedforward echo effect is created with tempo-synchronized delay.

Examples: Create and run the following m-file in MATLAB.

```matlab
% ECHOSYNC
% This script demonstrates one example to create a
% feedforward, tempo-synchronized echo effect
%
% See also CONVERTTEMPOSAMPLES


% Import our audio file
[x,Fs] = audioread('AcGtr.wav');

% Known tempo of recording
beatsPerMin = 102;  % units of beats per minute

% Calculate beats per second
beatsPerSec = beatsPerMin / 60; % 1 minute / 60 seconds

% Calculate # of seconds per beat
secPerBeat = 1/beatsPerSec;

% Note division
% 4 = whole, 2 = half, 1 = quarter, 0.5 = 8th, 0.25 = 16th
noteDiv = 0.5 ;
```

```
% Calculate delay time in seconds
timeSec = noteDiv * secPerBeat;


% Convert to units of samples
d = fix(timeSec * Fs); % round to nearest integer sample


b = 0.75; % amplitude of delay branch


% Total number of samples
N = length(x);
y = zeros(N,1);
% Index each element of our signal to create the output
for n = 1:N

    % When the sample number is less than the time delay
    % Avoid indexing a negative sample number
    if n < d+1
        % output = input
        y(n,1) = x(n,1);

    % Now add in the delayed signal
    else
        % output = input + delayed version of input
        % reduce relative amplitude of delay to 3/4
        y(n,1) = x(n,1) + b*x(n-d,1);
    end
end

sound(y,Fs); % Listen to the effect
```

Example 11.3: Script: echoSync.m

The feedforward echo in Figure 11.4 that is implemented in Example 11.3 creates a single repetition blended with the input signal.

11.5.2 Multi-Tap Echo

The **multi-tap echo** is one effect that creates multiple delayed repetitions of an input signal. This effect is implemented using a feedforward design with multiple delay lines in parallel.

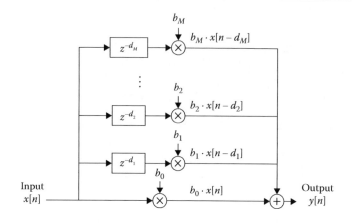

$$y[n] = b_0 \cdot x[n] + b_1 \cdot x[n-d_1] + b_2 \cdot x[n-d_2] + \ldots + b_M \cdot x[n-d_M]$$

Figure 11.5: Block diagram of multi-tap echo

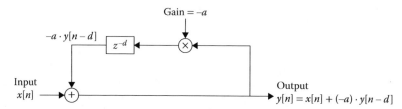

Figure 11.6: Block diagram of feedback echo

Each delay line can have a different delay time, $z^{-d_1}, \cdots, z^{-d_M}$ and a different relative amplitude, $b_0, \cdots, b_M$. A block diagram of this effect is shown in Figure 11.5.

11.6 Feedback Echo

A different method of creating an echo effect is the **feedback echo**. A block diagram for this type of system is shown in Figure 11.6. This system is described as a feedback echo because the output is fed to the input on a backward, delayed path. In this case, the input signal is added together with a delayed version of the output signal.

11.6.1 Feedback Gain Convention

As a convention, the gain of a feedback delay line in a block diagram is written as negative a. This convention is followed to remain consistent with MATLAB, and to allow the difference equation to be arranged as follows:

$$y[n] = x[n] + (-a) \cdot y[n-d]$$

$$y[n] + a \cdot y[n-d] = x[n]$$

In Example 11.4, consider the system in Figure 11.6 with difference equation $y[n] = x[n] + (-a) \cdot y[n-d]$. A feedback echo effect is created with tempo-synchronized delay.

Examples: Create and run the following m-file in MATLAB.

```matlab
% ECHOFEEDBACK
% This script demonstrates one example to create a
% feedback, tempo-synchronized echo effect
%
% See also ECHOSYNC

% Import our audio file
[x,Fs] = audioread('AcGtr.wav');

% Known tempo of recording
beatsPerMin = 102;   % units of beats per minute

% Calculate beats per second
beatsPerSec = beatsPerMin / 60; % 1 minute / 60 seconds

% Calculate # of seconds per beat
secPerBeat = 1/beatsPerSec;

% Note division
% 4 = whole, 2 = half, 1 = quarter, 0.5 = 8th, 0.25 = 16th
noteDiv = 0.5 ;

% Calculate delay time in seconds
timeSec = noteDiv * secPerBeat;

% Convert to units of samples
d = fix(timeSec * Fs); % round to nearest integer sample

a = -0.75; % amplitude of delay branch

% Index each element of our signal to create the output
N = length(x);
y = zeros(N,1);
for n = 1:N
```

```
% When the sample number is less than the time delay
% Avoid indexing a negative sample number
if n < d+1
    % output = input
    y(n,1) = x(n,1);

    % Now add in the delayed signal
else
    % output = input + delayed version of output
    % reduce relative amplitude of delay to 3/4
    y(n,1) = x(n,1) + (-a)*y(n-d,1);
    end
end

sound(y,Fs); % Listen to the effect
```

Example 11.4: Script: `echoFeedback.m`

11.6.2 Delayed Repetitions

An important difference between the feedforward echo and feedback echo is the number of delayed repetitions. For the feedforward delay the number of repetitions is equal to the number of delay lines with nonzero gain. For the feedback delay with nonzero gain, there is an infinite number of repetitions (theoretically).

Many times for practical purposes, the processing of a system with feedback delay is truncated over time to a finite length. This was performed in Example 11.4 such that the length of the output signal was identical to the length of the input signal. In other situations, the output signal is truncated after decreasing below an amplitude threshold.

11.6.3 System Stability

For Figure 11.6, consider three possibilities: $0 < |a| < 1, |a| = 1, |a| > 1$. First, when $0 < |a| < 1$, each feedback loop through the delay line decreases the amplitude of the output signal. Even after the input signal is finished, the output signal decays but never ends. Audio engineers call this type of system a **stable system** because the repetitions decay towards zero.

Second, when $|a| > 1$, each feedback loop through the delay line increases the amplitude of the output signal. Eventually, the amplitude of the output signal approaches $\pm\infty$. Audio engineers call this type of system an **unstable system** because the repetitions have no bound and can increase in amplitude indefinitely. Typically, an unstable system is undesirable.

Finally, when $|a| = 1$, the input signal repeats indefinitely without decreasing or increasing. Audio engineers call this type of system a **marginally stable system** because the repetitions do not decay to zero, but they do not increase towards infinity. Rather, the repetitions persist indefinitely at unity gain.

11.6.4 Combined Feedforward and Feedback Echo

It is possible to create systems with both feedforward and feedback echo. In this case, each feedforward delay line can have a separate delay time and amplitude from the feedback delay lines. In general, there can be multiple feedforward and multiple feedback delay lines. An example of a system with combined feedforward and feedback echo is shown in Figure 11.7. The difference equation for the effect is based on both the delayed input signal, $x[n - d_M]$, and the delayed output signal, $y[n - d_W]$.

Output Gain Convention

In Figure 11.7, there is not a gain multiplication block included on the output path. This is a common convention used in signal processing. If there is output gain, a_0, it is not used to directly scale the output, $y[n]$. Rather, a_0 is distributed to the other gain terms to maintain unity gain for the output in the difference equation.

$$a_0 \cdot y[n] + a_W \cdot y[n - d_W] = b_0 \cdot x[n] + b_M \cdot x[n - d_M]$$

$$y[n] + \frac{a_W}{a_0} \cdot y[n - d_W] = \frac{b_0}{a_0} \cdot x[n] + \frac{b_M}{a_0} \cdot x[n - d_M]$$

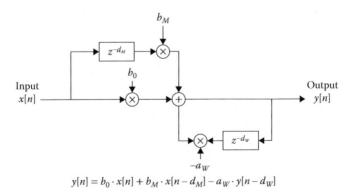

$$y[n] = b_0 \cdot x[n] + b_M \cdot x[n - d_M] - a_W \cdot y[n - d_W]$$

Figure 11.7: Combined feedforward and feedback delay block diagram

11.6.5 Stereo Echo

Audio engineers use echo in effects with a stereo output. Even if a mono input signal is used, the result is a two-channel signal with delayed repetitions occurring across the stereo field.

There are many different variations of the stereo echo. One foundational approach uses a feedforward, multi-tap echo with the individual taps sent separately to the left and right channels. An example of a block diagram for this type of stereo echo is shown in Figure 11.8.

11.6.6 Ping-Pong Echo

The use of feedback provides another approach to creating a stereo delay. Multiple repetitions are achieved at different locations across the stereo field by routing the output of a delay block back to the input of a delay block.

One noteworthy version of this effect is the **ping-pong echo**. For this effect, repetitions of a signal alternate from the left and right channels in a pattern across the stereo field. It is created by routing the output of the delay block for the left channel to the input of the delay block for the right channel, and vice versa. The decay of the repetitions is controlled by a feedback gain, g. A block diagram of the ping-pong echo effect is shown in Figure 11.9.

11.7 Impulse Response

An important way to both analyze and describe a system with memory is by taking an **impulse response**. This process involves measuring and representing the final output of a system due to an impulse input signal.

A digital impulse signal has an amplitude of zero for all time samples except at the first time sample, where the amplitude is 1. Another name for the digital impulse signal is the

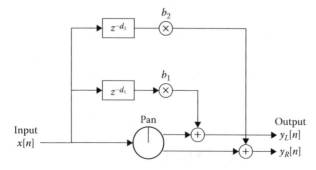

Figure 11.8: Block diagram of feedforward stereo echo

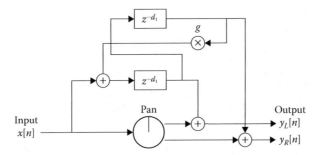

Figure 11.9: Block diagram of ping-pong echo

Kronecker delta function, $\delta[n]$. As shown in Section 7.3.6, this type of signal can also be used to synthesize an impulse train signal.

Conceptually, a system's impulse response represents how an individual sample of a signal is processed by an effect. The result is a sequence of amplitude values indicating how the impulse is delayed over time and the level of the impulse when delayed. Because all digital signals are comprised of a sequence of impulses, the idea is "if the response of the system is known for one sample, then the response of the system can be inferred for all samples in any signal." Therefore, the impulse response can be used to determine how a system responds to any input signal (given the assumptions described in Section 11.9).

11.7.1 Finite Impulse Response Systems

If a system with memory uses only feedforward delay lines, then the impulse response of the system has a finite length. Therefore, another term to describe a feedforward system is a **finite impulse response** (FIR) system. Regardless of the number of delay lines, as long as the time of the delay with the greatest duration is less than infinity, then the system's response has a finite length. In Example 11.5, the impulse response of an FIR system is measured. In Figure 11.10, the input impulse signal is compared to the output impulse response.

Examples: Create and run the following m-file in MATLAB.

```
% IMPFIR
% This script demonstrates one example to measure
% the impulse response of an FIR system
%
% See also IMPIIR

clear;clc;close all;
Fs = 48000;
```

```
N = Fs*2;
% Synthesize the impulse signal
imp = zeros(N,1); % 2 second long signal of 0's
imp(1,1) = 1;  % Change the first sample = 1

d1 = 0.5*Fs;  % 1/2 second delay
b1 = 0.7; % Gain of first delay line

d2 = 1.5*Fs;  % 3/2 second delay
b2 = 0.5;     % Gain of second delay line

% Zero-pad the beginning of the signal for indexing
% based on the maximum delay time
pad = zeros(d2,1);
impPad = [pad;imp];

out = zeros(N,1);
% Index each element of our signal to create the output
for n = 1:N
    index= n+d2;
    out(n,1) = impPad(index,1) ...
        + b1*impPad(index-d1,1) + b2*impPad(index-d2,1);
end

t = [0:N-1]/Fs;
subplot(1,2,1);
stem(t,imp); % Plot the impulse response
axis([-0.1 2 -0.1 1.1]);
xlabel('Time (sec.)');
title('Input Impulse');
subplot(1,2,2);
stem(t,out); % Plot the impulse response
axis([-0.1 2 -0.1 1.1]);
xlabel('Time (sec.)');
title('Output Impulse Response');

% Save the impulse response for future use
audiowrite('impResp.wav',out,Fs);
```

Example 11.5: Script: `impFIR.m`

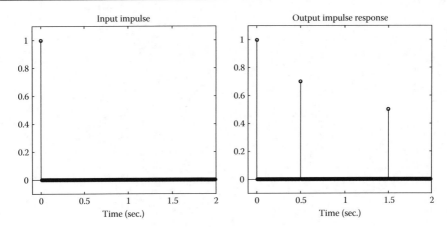

Figure 11.10: Impulse response of FIR system

11.7.2 Infinite Impulse Response Systems

If a system with memory has at least one feedback delay line, then the impulse response of the system has an infinite length. Therefore, another term to describe a feedback system is an **infinite impulse response** (IIR) system. In Example 11.6, the impulse response of an IIR system is approximated by truncating the output to a finite length after the amplitude decreased to a negligible amount. In Figure 11.11, the input impulse signal is compared to the output impulse response.

Examples: Create and run the following m-file in MATLAB.

```
% IMPFIR
% This script demonstrates one example to approximate
% the impulse response of an IIR system
%
% See also IMPFIR

clear;clc;close all;
Fs = 48000; Ts = 1/Fs;
N = Fs*2;    % Number of samples
% Synthesize the impulse signal
imp = zeros(N,1); % 2 second long signal of 0's
imp(1,1) = 1;  % Change the first sample = 1

d1 = 0.5*Fs;   % 1/2 second delay
a1 = -0.7; % Gain of feedback delay line
```

```
% Index each element of our signal to create the output
for n = 1:d1
    out(n,1) = imp(n,1); % Initially there is no delay
end
for n = d1+1:Fs*2        % Then there is signal + delay
    out(n,1) = imp(n,1) + a1*out(n-d1);
end
for n = Fs*2+1:Fs*10     % Finally there is only delay
    out(n,1) = a1*out(n-d1);  % After input finished
end
t = [0:N-1]*Ts;
subplot(1,2,1);
stem(t,imp); % Plot the impulse response
axis([-0.1 2 -0.1 1.1]);
xlabel('Time (sec.)');
title('Input Impulse');
subplot(1,2,2);
t = [0:length(out)-1]*Ts;
stem(t,out); % Plot the impulse response
axis([-0.1 10 -1.1 1.1]);
xlabel('Time (sec.)');
title('Output Impulse Response');
```

Example 11.6: Script: impIIR.m

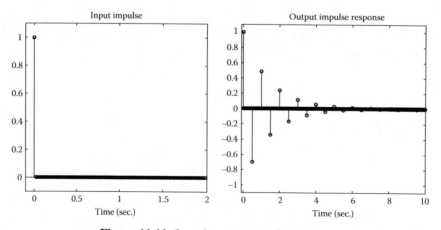

Figure 11.11: Impulse response of IIR system

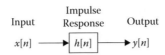

Figure 11.12: Block diagram of a generalized linear system

11.8 Convolution

If an impulse response is a measurement and description of a system's behavior, then **convolution** is the mathematical operation of processing a signal going through the system by using the impulse response. Essentially, each input sample, $x[n]$, is independently sent though the system, $h[n]$, to contribute to the output signal, $y[n]$, at different times. Therefore, the output signal is created by summing together a sequence of impulse responses due to each input sample. In Figure 11.12, a block diagram is shown for the relationship between the input and output of a linear system. The mathematical operation for convolution is written as $y[n] = x[n] * h[n]$, where the symbol $*$ is used to denote convolution, and not multiplication.

11.8.1 MATLAB Convolution Function

A built-in MATLAB function is available to perform convolution: $y=\text{conv}(x,h)$. The input variables to the function are a signal, x, and a system, h, which are both a sequence of samples. The output variable, y, is a sequence of samples that represents the processed signal.

In Example 11.7, a sine-wave input signal is convolved with the impulse response of the FIR system in Example 11.5. A comparison of the input signal, impulse response, and output signal is shown in Figure 11.13.

Examples: Create and run the following m-file in MATLAB.

```
% convolutionExample.m
% This script demonstrates the MATLAB convolution
% function - "y = conv(x,h)"
% The example demonstrated is with a single cycle of a
% sine wave. When the sine wave is convolved with the
% impulse response for an echo effect, the output
% signal has delayed copies of the sine wave at
% different amplitudes at different times.
%
% See also CONV

clear; clc; close all;
% Import previously saved IR
```

```
[h,Fs] = audioread('impResp.wav');

% Synthesize input signal
x = zeros(2*Fs,1); % 2 second long input signal
f = 4;
t = [0:Fs*0.25 - 1]/Fs;
x(1:Fs*0.25) = sin(2*pi*f*t);

% Perform convolution
y = conv(x,h);

% Plot signals
xAxis = (0:length(h)-1)/Fs;
subplot(3,1,1);
plot(xAxis,x); % Plot the impulse response
axis([-0.1 2 -1.1 1.1]);
xlabel('Time (sec.)');
title('Input Signal - x[n]');

subplot(3,1,2);
stem(xAxis,h); % Plot the impulse response
axis([-0.1 2 -0.1 1.1]);
xlabel('Time (sec.)');
title('Impulse Response - h[n]');

subplot(3,1,3);
plot(xAxis,y(1:2*Fs)); % Plot the impulse response
axis([-0.1 2 -1.1 1.1]);
xlabel('Time (sec.)');
title('Output Signal - y[n]');
```

Example 11.7: Script: `convolutionExample.m`

11.8.2 *Mathematical Equation for Convolution*

The following equation for convolution can be used to determine the amplitude of the output signal, $y[n]$, at any sample number, $n = 1, 2, 3, ..$:

$$y[n] = x[n] * h[n] = \sum_{m=1}^{M} x[n-m+1] \cdot h[m], \text{ where } M = \texttt{length(h)} \qquad (11.1)$$

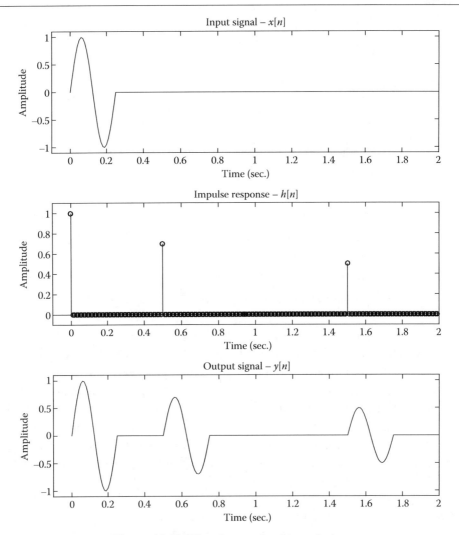

Figure 11.13: Visual example of convolution

Conceptually, the input signal is fed through the system by starting with the first sample and iterating to the end. This is accomplished by time-reversing the input signal then performing an element-wise multiplication with the processing system as the signal passes through.

11.8.3 Convolution Reverberation

In addition to using convolution to create echo effects, it is also used to create digital reverberation. Intuitively, reverb is the sound of many closely spaced echoes, or signal repetitions. Convolution can create time delays whether the repetitions are sparsely or densely spaced.

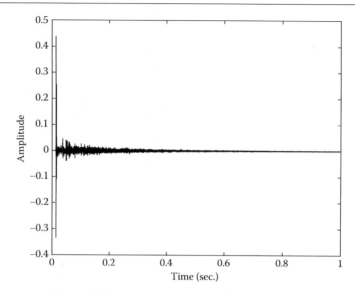

Figure 11.14: Impulse response of acoustic reverb

To create convolution reverb, a signal is convolved with an impulse response. Several types of reverb effects can be approximated by an impulse response including an acoustic space, mechanical plate and spring reverb, as well as digital reverb. Stereo convolution reverb is created by measuring two different impulses: one to be convolved with the input signal to create the left channel of the output signal, and another to be convolved with the input signal to create the right channel of the output signal. In Example 11.8, stereo convolution reverb is created by using a two-channel impulse response measured at a recording studio in Nashville, Tennessee. In Figure 11.14, the impulse response is plotted for the left channel.

Examples: Create and run the following m-file in MATLAB.

```
% REVERBCONV
% This script demonstrates the process to create
% a stereo convolution reverb by using a two-channel
% impulse response. This impulse response is based
% on a measurement of a recording studio
% in Nashville, Tennessee.

clear;clc;close all;

% Import sound file and IR measurement
[x,Fs] = audioread('AcGtr.wav'); % Mono signal
[h] = audioread('reverbIR.wav'); % Stereo IR
```

```
% Visualize one channel of the impulse response
plot(h(:,1));

% Perform convolution
yLeft = conv(x,h(:,1));
yRight = conv(x,h(:,2));


y = [yLeft,yRight];


sound(y,Fs);
```

Example 11.8: Script: reverbConv.m

Besides using convolution, another approach to simulating the sound of an acoustic space is algorithmic reverb, discussed in Chapter 16.

11.9 Necessary Requirements for Modeling a System with an Impulse Response and Convolution

There are two requirements for modeling a system (digital, electronic, acoustic, etc.) using an impulse response and convolution. The system must be **linear** over the operating range modeled and it must be **time-invariant** (LTI).

11.9.1 Linearity

Many audio signal processing systems can be modeled as linear systems. A linear system processes the amplitude of the input signal by a linear combination of a signal's samples. This combination involves *multiplying* the signal's amplitude by scalar values and *adding* together the result with previous samples.

Conceptually, linear systems change the gain of a signal by the same amount regardless of the amplitude of the input signal. Conversely, nonlinear systems change the gain of a signal by one amount when the amplitude is low and a different amount when the amplitude is high.

Examples of Nonlinear Systems

There are many examples of nonlinear systems in audio. Vacuum tubes are known for amplifying signals such that when a signal's amplitude is high enough, the gain change distorts. Analog tape saturation is similar because if the input signal has a high amplitude, then the tape domains distort. Nonlinear digital systems are discussed in Chapter 10.

A dynamic-range compressor is a nonlinear system because the signal amplitude is not changed when it is below the threshold, but the amplitude is attenuated when it is above the threshold. Compressors are presented in Chapter 18.

Nonlinear systems cannot be modeled with a single impulse response because convolution applies the same processing to each sample in a signal regardless of whether the amplitude is low or high.

11.9.2 Time Invariance

A system that is time-invariant processes a signal in the same way for all time. In other words, the system is static and does not change. If the system's response is contant, it can be described as *stationary*. An impulse response represents a snap-shot of a system, which is then applied in the same way to all samples of an input signal. The system's performance is independent of how it has performed previously and how it performs in the future. In Chapter 12, spectral processors are discussed as another example of LTI systems.

Examples of Time-Variant Systems

There are many examples of time-variant systems in audio. Modulation effects such as chorus, flanging, phasing, ring modulation, etc. are time-variant because the system's delay time changes over time. In Chapter 15, time-variant, modulated delay effects are presented. Dynamic-range processors are also a type of time-variant system because the gain reduction at any point in time depends on the past gain reduction due to attack and release parameters. In Chapter 18, compressors, limiters, expanders, and gates are presented.

Chapter 12

Finite Impulse Response Filters *Systems with Memory: Spectral Processors (Part 1)*

12.1 Introduction: Spectral Processors

This chapter introduces the use of time delays in systems to create spectral processors. These audio effects include various types of filters for equalizers. In Chapter 11, relatively long time delays were used to create audible echo effects. Here, relatively short time delays are used to create spectral processors.

Spectral processors change the characteristics of a signal in a frequency-dependent manner. In other words, one frequency is changed differently than another frequency. Spectral processors are described based on how the amplitude of a signal is changed in a frequency dependent manner. Examples of spectral processors are low-pass filters, high-pass filters, band-pass filters, band-stop filters, comb filters, peaking filters, shelving filters, and all-pass filters.

Filters using a feedforward design are presented in this chapter. Subsequently, filters using a feedback design are discussed in Chapter 13.

12.1.1 Filter Characteristics

There are several important characteristics used to describe the performance of different types of spectral processors (Figure 12.1). The **pass band** of a filter is the frequency range where a signal passes through the system (relatively) unchanged. The **stop band** of a filter is the frequency range where a signal is attenuated by the system. The **transition band** of a filter is the frequency range between the pass band and the stop band. The **cutoff frequency** (f_c) is the point in the spectrum when a signal is changed by 3 dB.

12.1.2 Visualizing the Spectral Response of a Filter

There is a built-in function for MATLAB that can be used to perform a spectral analysis of a system. The function is called `freqz`. The many uses of this function are demonstrated throughout the remainder of this chapter.

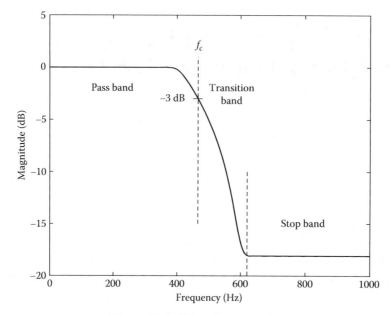

Figure 12.1: Filter characteristics

12.1.3 Units of Time Delay

For echo effects, time units of seconds and samples are used to describe delay length depending on context (Section 11.3). This is also true for spectral effects. Additionally, when a short delay is introduced, it is helpful to consider time units relative to the length of a cycle for a given frequency. This can be done as a fraction of a cycle, as well as in units of degrees or radians for a phase shift. Examples of the various units used to describe a time delay are shown in Figure 12.2.

12.2 Basic Feedforward Filters

The fundamental approach to creating a spectral processor is based on combining a signal with time-delayed versions of the signal. As a foundation for more complex processors, a basic low-pass filter and high-pass filter are initially considered. The basic design uses a single, parallel delay line with a delay of one sample. Then, the complexity of the basic filter can be increased by exploring other spectral processors.

12.2.1 Low-Pass Filter

A **low-pass filter** (LPF) is a spectral processor that attenuates the amplitude of high frequencies while letting low frequencies pass through. In Figure 12.3, an example of an LPF

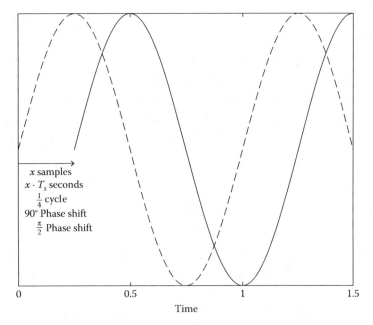

Figure 12.2: Various units of a time delay

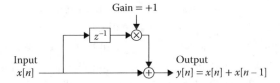

Figure 12.3: Block diagram of LPF using one-sample delay

is shown. Here, the unprocessed input signal is added together with an input signal that has been delayed by one sample.

Basic LPF Impulse Response

The impulse response of the system in Figure 12.3 is $h = [1\ 1]$. The frequency response of the system can be visualized by using the command $freqz(h)$.

The frequency response is displayed as two plots. The top plot is called the **magnitude response**, or amplitude response, for the system. It represents how the amplitude of a signal changes at a particular frequency. The bottom plot is called the **phase response** for the system. It represents the phase shift on an input signal at a particular frequency.

The horizontal axis of the frequency response spans a range from 0–1 ($\times \pi \frac{radians}{sample}$). The units are in *normalized frequency*, spanning from 0 Hz to the Nyquist frequency. In other words, the units could be considered as *relative frequency*, dependent on the sampling rate of the input signal.

Consider a signal with a sampling rate of 48,000 $\frac{samples}{second}$. If it is processed by this LPF, the magnitude and phase response of this plot at a frequency of 0.1 represents how the input signal is changed at the frequency, $24,000 \cdot 0.1 = 2400$ Hz. For a signal with a sampling rate of 44,100 $\frac{samples}{second}$, the magnitude and phase response of this plot at a frequency of 0.1 represents how the input signal is changed at the frequency, $22,050 \cdot 0.1 = 2205$ Hz.

Sine Wave Analysis

One method to analyze the behavior of a system is to use a sine wave as an input signal. In this case, the behavior of a system is analyzed at an individual frequency.

Consider the LPF in Figure 12.3 for a 1 Hz sine wave input signal. As an example, assume a sampling rate of 48,000 $\frac{samples}{second}$. The number of samples per cycle for a 1 Hz sine wave is:

$$\frac{48,000 \; samples}{1 \; second} \cdot \frac{1 \; second}{1 \; cycle} = \frac{48,000 \; samples}{1 \; cycle}$$

In this system, the input signal is added together with the input signal after a one-sample delay. For an input signal which completes a cycle after 48,000 samples, there is a negligible phase difference between the delayed signal and the unprocessed signal. In other words, this system creates *constructive interference* for the 1 Hz input signal (i.e., low frequencies). When the two signal paths are combined at the output, the amplitude of the input signal essentially doubles, or increases by $\sim$6 dB.

As a contrast, consider a high-frequency sine wave as the input signal. For this sampling rate, the Nyquist frequency is 24,000 Hz. The number of samples per cycle for a 24,000 Hz sine wave is:

$$\frac{48,000 \; samples}{1 \; second} \cdot \frac{1 \; second}{24,000 \; cycles} = \frac{2 \; samples}{1 \; cycle}$$

For an input signal that completes a cycle after two samples, the one-sample delay is the equivalent of a $\frac{1}{2}$-cycle delay. In other words, this system creates a 180° phase difference between the input signal and the delayed signal. Therefore, *destructive interference* occurs for the 24,000 Hz input signal (i.e., high frequencies). The amplitude at the output is attenuated.

The underlying concept for spectral processing is that a constant time delay introduces a different amount of phase shift for different frequencies. The difference in phase shift creates constructive interference for some frequencies while other frequencies have destructive interference to varying degrees.

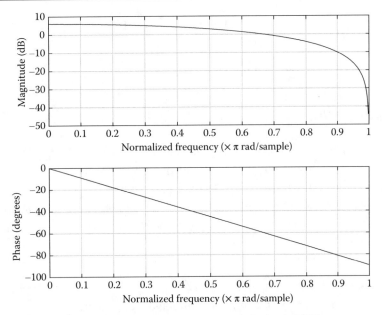

Figure 12.4: Frequency response of a basic LPF

The phase difference, ϕ, for the delayed signal compared to the unprocessed signal at any frequency, f, can be determined by:

$$\phi \; \frac{radians}{1 \; sample} = \frac{2 * \pi \; radians}{cycle} \cdot \frac{f \; cycles}{second} \cdot \frac{1 \; seconds}{48,000 \; samples}$$

The difference equation from Figure 12.3 can be written based on a sine wave input signal and phase difference. Using a trigonometric identity, the amplitude, A, and phase, ϕ, of the system's output for a given frequency is calculated. This result can be used to determine the frequency response in Figure 12.4.

$$sin(x) + sin(x + \phi)$$

$$Note: \; sin(u) + sin(v) = 2 * sin\left(\frac{u+v}{2}\right) * cos\left(\frac{u-v}{2}\right)$$

$$sin(x) + sin(x + \phi) = 2 * sin\left(\frac{x+x+\phi}{2}\right) * cos\left(\frac{x-x-\phi}{2}\right)$$

$$sin(x) + sin(x + \phi) = 2 * sin\left(x + \frac{\phi}{2}\right) * cos\left(\frac{-\phi}{2}\right)$$

$$Let:\ A = 2 * cos\left(\frac{-\phi}{2}\right)$$

$$sin(x) + sin(x+\phi) = A * sin\left(x+\frac{\phi}{2}\right)$$

12.2.2 High-Pass Filter

A **high-pass filter** (HPF) is a spectral processor that attenuates the amplitude of low frequencies while letting high frequencies pass through. One example of an HPF is shown in Figure 12.5. This block diagram is similar to the LPF in Figure 12.3 with the exception of the gain of −1 on the delay line. Therefore, the input signal is delayed by one sample, as well as inverted in polarity.

Basic HPF Impulse Response

The impulse response of the system in Figure 12.5 is h = [1 -1]. The frequency response of the system is shown in Figure 12.6 using the command freqz(h).

Sine Wave Analysis

A sine wave analysis similar to that in Section 12.2.1 can be used to demonstrate the frequency response of the HPF. By inverting the polarity of the signal on the delay line, the opposite type of constructive/destructive interference occurs compared to the LPF.

The one-sample delay creates the same phase difference for the input signal. By changing the gain on the delay line, the delayed signal is subtracted from the unprocessed input signal.

The difference equation from Figure 12.5 can be written based on a sine wave input signal and phase difference. Using a trigonometric identity, the amplitude and phase of the system's output is discovered. This result can be used to determine the frequency response in Figure 12.6.

$$sin(x) - sin(x+\phi)$$

$$Note:\ sin(u) - sin(v) = 2 * sin\left(\frac{u-v}{2}\right) * cos\left(\frac{u+v}{2}\right)$$

$$sin(x) - sin(x+\phi) = 2 * sin\left(\frac{x-x-\phi}{2}\right) * cos\left(\frac{x+x+\phi}{2}\right)$$

$$sin(x) - sin(x+\phi) = 2 * sin\left(-\frac{\phi}{2}\right) * cos\left(\frac{x+x+\phi}{2}\right)$$

$$Let:\ A = 2 * sin\left(\frac{-\phi}{2}\right)$$

$$sin(x) - sin(x+\phi) = A * cos\left(x+\frac{\phi}{2}\right)$$

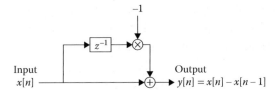

Figure 12.5: Block diagram of HPF using 1-sample delay

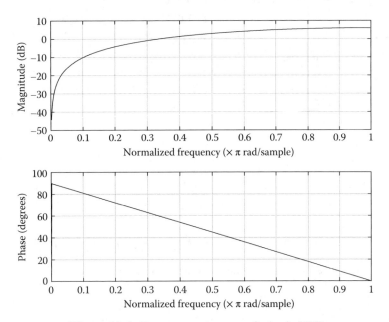

Figure 12.6: Frequency response of a basic HPF

Filter Order

Filters that only use a single sample of delay are called **first-order systems**. Filters that use two samples of delay are called **second-order systems**. This naming convention applies for higher-order systems as well.

12.3 Additional Feedforward Filters

Other types of filters can be created by changing the delay time of the feedforward delay line. A comb filter and a band-pass filter can be created using two (or more) samples of delay.

12.3.1 Comb Filter

A **comb filter** is a spectral processor that attenuates one or more frequencies within the bandwidth of a system. When visualizing the magnitude response of the system, notches occur spaced across the spectrum. Therefore, when a signal is procesed by this system, it is as if a comb is carving through the spectrum to attenuate some frequencies. In Figure 12.7, one example of a comb filter is shown that creates a notch at half the Nyquist frequency.

Comb Filter Impulse Response

The impulse response of the comb filter in Figure 12.7 is h = [1 0 1]. For this system, the impulse response is one at h(1) because of the unprocessed input signal. The impulse response is zero at h(2) because there is not a feedforward branch with one sample of delay. Therefore, the impulse input signal does not occur in the output after one sample of delay. Finally, the impulse response is one at h(3) because of the branch with two samples of delay. The frequency response of the system is shown in Figure 12.8.

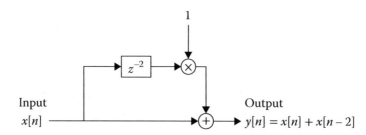

Figure 12.7: Block diagram of comb filter using two-sample delay

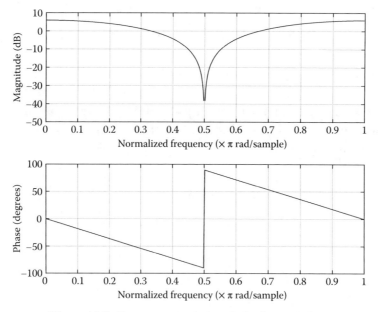

Figure 12.8: Frequency response of a basic comb filter

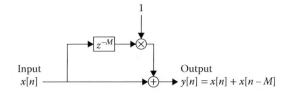

Figure 12.9: Block diagram of comb filter using M-sample delay

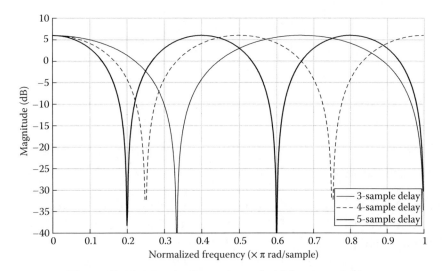

Figure 12.10: Magnitude response of additional comb filters

Additional Comb Filters

By changing the number of samples of delay, different comb filters can be created. In Figure 12.9, a generalized block diagram is shown for an arbitrary number of samples of delay, M. The resulting magnitude response is shown in Figure 12.10 for several examples.

The following relationship can be used for the whole numbers, $m = 0, 1, 2, \ldots, M$ to determine if a frequency, f, has constructive or destructive interference from the comb filter created by Figure 12.9 with M samples of delay.

$$f = \frac{m}{M} * Nyquist \tag{12.1}$$

Even numbers, $m = 0, 2, 4, \ldots, M$, can be used to determine the relative frequencies where constructive interference occurs. Odd numbers, $m = 1, 3, 5, \ldots, M$, can be used to determine the relative frequencies where destructive interference occurs.

Sine Wave Analysis

The amplitude and phase response of the comb filter at a given frequency can be determined by using the trigonometric identity for the sum of sines. Since the filter does not have one sample of delay, but can have M samples of delay, the phase shift, ϕ_M, of the delay should be determined. If there are ϕ radians for one sample of delay, then there are $M \cdot \phi$ radians for M samples of delay. Therefore:

$$\frac{\phi\ radians}{1\ sample} \rightarrow \frac{M \cdot \phi\ radians}{M\ samples} = \phi_M$$

$$sin(x) + sin(x + \phi_M) = A * sin\left(x + \frac{\phi_M}{2}\right)$$

$$Let:\ A = 2 * cos\left(\frac{-\phi_M}{2}\right)$$

12.3.2 Band-Pass Filter

A **band-pass filter** (BPF) is a system that attenuates the amplitude of low frequencies and high frequencies while letting a band of frequencies pass through unattenuated. One example of a BPF is shown in Figure 12.11.

BPF Impulse Response

The impulse response of the BPF in Figure 12.11 is h = [1 0 -1]. For this system, the impulse response is negative one at h(3). The frequency response of the system is shown in Figure 12.12.

12.3.3 Feed-forward Comb Filter with Gain = –1

Another modification of the feedforward filter is to increase the number of samples of delay while using a negative gain for the delay line. An example of this system is shown in Figure 12.13.

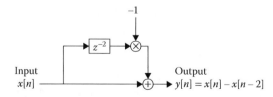

Figure 12.11: Block diagram of BPF using 2-sample delay

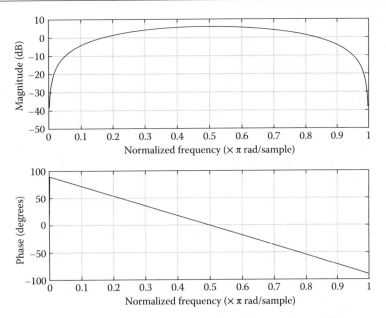

Figure 12.12: Frequency response of a basic band-pass filter

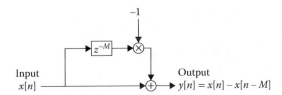

Figure 12.13: Block diagram of feed-forward comb filter with gain $= -1$

As was demonstrated in Section 12.2.1 with an LPF and in Section 12.2.2 with an HPF, by changing the gain on a delay from $+1$ to -1, the frequencies with constructive and destructive interference are reversed. This is shown in Figure 12.14 for the comb filter in comparison to Figure 12.10.

Therefore, a similar relationship can be used for the whole numbers, $m = 0, 1, 2, \ldots, M$, to determine if a frequency, f, has constructive or destructive interference from the comb filter created by Figure 12.13 with M samples of delay.

$$f = \frac{m}{M} * Nyquist$$

In this case, odd numbers, $m = 1, 3, 5, \ldots, M$, can be used to determine the relative frequencies where constructive interference occurs. Even numbers, $m = 0, 2, 4, \ldots, M$, can be used to determine the relative frequencies where destructive interference occurs.

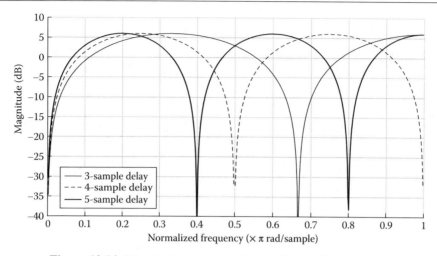

Figure 12.14: Magnitude response of comb filters with gain $= -1$

12.4 Changing the Relative Gain of a Filter

There are several locations in the diagram of a filter where the gain of a signal can be changed. Furthermore, these gain parameters of a filter can be set to gain values other than ± 1. By changing the gain, the resulting frequency response of the system is changed. Therefore, the gain parameters represent additional *degrees of freedom* for modifying system behavior.

For notation, an arbitrary gain scalar on the direct, dry path of the filter is labeled b_0. The gain scalar on the delay line with one sample of delay is labeled b_1. In general, the gain scalar on the delay line with M samples of delay is labeled b_M.

12.4.1 Changing the Gain of the Delay Line

One way to change the frequency response of a filter is by altering the relative gain between the dry, unprocessed path and the delay line. When the magnitude of the amplitude for each path is equal, the maximum constructive and destructive interference occurs for frequencies given in Equation 12.1. When the magnitude of the amplitude for each path is different, less constructive and destructive interference occurs. Therefore, the maximum and minimum extremes of the system's magnitude response are diminished.

As an example, consider the system in Figure 12.15 where the gain $b_0 = 1$ and $b_3 = \beta$ for $\beta \in [0, 1]$. Therefore, the impulse response of the system is $h = [1 \ 0 \ 0 \ \beta]$. The frequency response of the system for various values of β is shown in Figure 12.16.

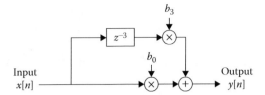

Figure 12.15: Block diagram of comb filter with arbitrary delay line gain

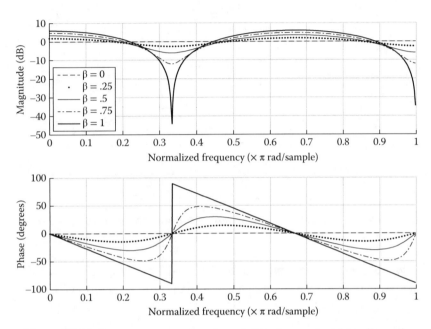

Figure 12.16: Frequency response of comb filter by varying delay line gain

As the amplitude, β, of the delay line increases from 0 to 1, it can be considered as blending in the frequency-dependent interference whether constructive or destructive. If β spans a range from 0 to -1, a similar blending occurs for the frequency response shown in Figure 12.13.

If the relative amplitude, $|\beta|$, for the delay line is greater than the dry, unprocessed path ($b_3 > b_0$), then the overall system gain is increased. However, the relative amplitude response is identical to the system for $b_3 = \frac{1}{\beta}$. In Figure 12.17, the frequency response is shown for the systems [1 0 0 0.5] and [1 0 0 2]. For this example, $\beta = \frac{1}{2}$ and $\beta = 2$. The only difference in the amplitude response is the overall system gain.

In general, it is the ratio between b_0 and b_3 that determines the relative amplitude response. Therefore, all pairs of b_0 and b_3 with ratio $R = \frac{b_0}{b_3}$ have a similar amplitude response. Additionally, the pairs of b_0 and b_3 that result with the ratio $\frac{b_0}{b_3} = \frac{1}{R}$ also have a similar

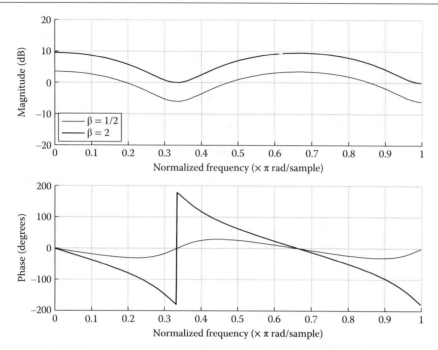

Figure 12.17: Frequency response of feedforward filter with $\beta > 1$

amplitude response. The only difference in the amplitude responses is the overall system gain. The maximum amplitude of the frequency response can be determined by $20 \cdot log_{10}(|b_0| + |b_3|)$.

Although the amplitude responses are similar in this situation, the phase responses are different. This relationship is exploited for the creation of other types of filters (Section 13.3.1).

Generalized Sine Wave Analysis

Consider a feedforward comb filter (Figure 12.18) with arbitrary gains, b_0 and b_M, as well as delay length, M. The amplitude response, A, and phase response, φ, can be determined for any frequency using a sine wave analysis. The following trigonometric identity describes the result:

$$b_0 \cdot sin(x) + b_M \cdot sin(x + \phi_M) = A \cdot sin(x + \varphi)$$

$$A = \sqrt{(b_0)^2 + (b_M)^2 + 2 \cdot b_0 \cdot b_M \cdot cos(\phi_M)}$$

$$\varphi = atan2\left(b_M \cdot sin(\phi_M), b_0 + (b_M \cdot cos(\phi_M))\right)$$

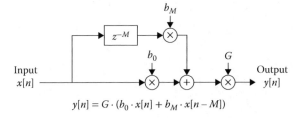

$$y[n] = G \cdot (b_0 \cdot x[n] + b_M \cdot x[n - M])$$

Figure 12.18: Block diagram of feedforward filter with normalized gain

12.4.2 Normalizing the Overall System Gain

As a convention, it is common for filters to be designed to have a pass band of unity gain, 0 dB. Then, the stop band is attenuated in amplitude relative to unity gain. Therefore, it is common to normalize the overall system gain. A scaling factor, G, can be included at the output of the filter for normalization where:

$$G = \frac{1}{|b_0| + |b_M|} \tag{12.2}$$

The gain parameters of the system are selected by convention such that only the desired spectral processing is contained within the filter effect. Any additional change in amplitude to the signal can be performed independent of the filter. The impulse response of the filter in Figure 12.18 with $M = 3$ is h = [G*b0, 0 , 0 , G*b3].

12.5 Generalized Finite Impulse Response Filters

In previous sections, a basic feedforward filter was presented followed by an analysis of changing the delay time and amplitude of the system. It was demonstrated how to use the parameters of the system to create several filters with various frequency responses. The feedforward filter can be futher generalized by including additional parallel delay lines, each with their own amplitude. The block diagram for the generalized FIR system is shown in Figure 12.19. This design is similar to the multi-tap echo effect described in Section 11.5.2.

By considering Equation 12.2, the scaling factor, G, can also be generalized by summing the amplitude of all delay lines. Therefore, the overall system gain can be normalized by using:

$$G = \frac{1}{\displaystyle\sum_{m=0}^{M} |b_m|}$$

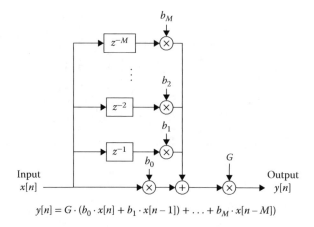

$$y[n] = G \cdot (b_0 \cdot x[n] + b_1 \cdot x[n-1]) + \ldots + b_M \cdot x[n-M])$$

Figure 12.19: Block diagram of generalized FIR filter

The impulse response of the generalized FIR filter can be written:

```
h = G * [b0 , b1 , b2 , ... , bM],
```

where b_0 is the gain for the input without delay, b_1 is the gain for the input delayed by one sample, etc., and b_M is the gain for the input delayed by M samples.

12.5.1 MATLAB FIR Filter Functions

There are several MATLAB functions that can be used to create filters. The following two functions can be used to create FIR filters.

MATLAB fir1 *Function*

The first filter design function is fir1. It can be used to create basic filter types including the LPF, HPF, BPF, and band-stop filter. The syntax to call this function is:

```
h = fir1(m,Wn,ftype);
```

- Input Variables
    - m : filter order, length of impulse response
    - Wn : normalized cutoff frequency
    - ftype : filter type ('low', 'high', 'bandpass', 'stop')
- Output Variable
    - h : output impulse response

In Example 12.1, this use of this function is demonstrated to create four types of filters. The frequency response of the filters are shown in Figures 12.20 to 12.24.

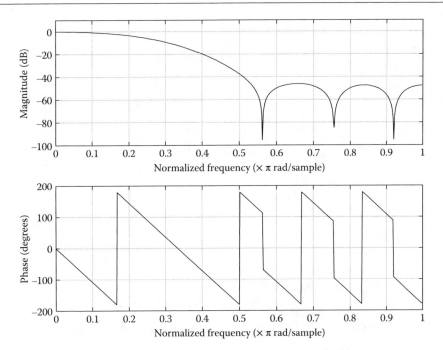

Figure 12.20: Frequency response of LPF created with fir1

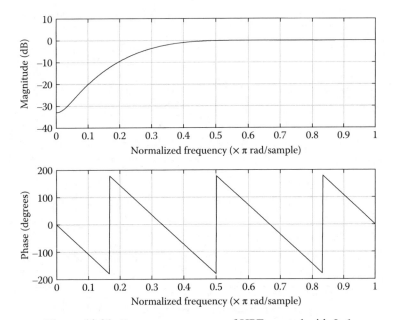

Figure 12.21: Frequency response of HPF created with fir1

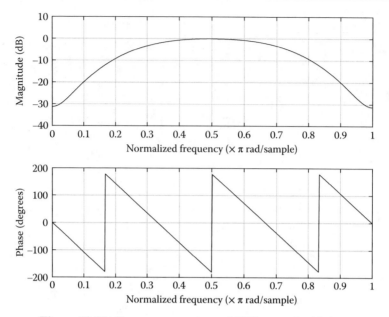

Figure 12.22: Frequency response of BPF created with fir1

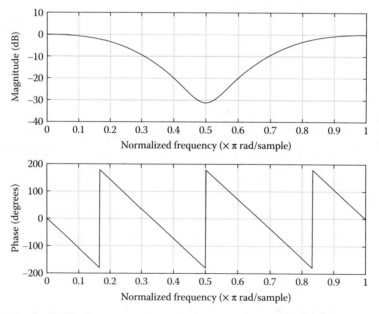

Figure 12.23: Frequency response of band-stop filter created with fir1

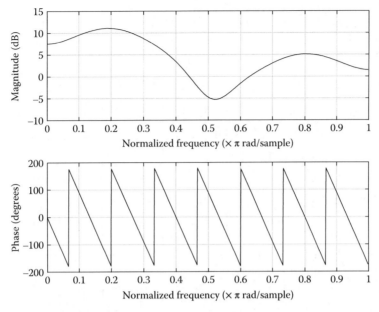

Figure 12.24: Frequency response of filter created with fir2

Examples: Create and run the following m-file in MATLAB.

```
% FIR1EXAMPLE
% This script demonstrates the various uses of
% the "fir1" filter function
%
% See also FIR1

% Declare initial parameters for LPF and HPF
n = 12; % order of filter = # of delay lines + 1
f = 6000; % cutoff frequency (Hz)
Fs = 48000; nyq = Fs/2;
Wn = f/nyq;

% Syntax for low-pass filter
h_lpf = fir1(n,Wn); % optionally, h = fir1(n,Wn,'low');

% Syntax for high-pass filter
h_hpf = fir1(n,Wn,'high');
```

```
% Declare second frequency for BPF and BSF
f2 = 18000;
Wn2 = f2/nyq;



% Syntax for band-pass filter
h_bpf = fir1(n,[Wn Wn2]);

% Syntax for band-stop filter
h_bsf = fir1(n,[Wn Wn2],'stop');



% Plots
figure(1); freqz(h_lpf); % Low-pass filter
figure(2); freqz(h_hpf); % High-pass filter
figure(3); freqz(h_bpf); % Band-pass filter
figure(4); freqz(h_bsf); % Band-stop filter
```

Example 12.1: Script: fir1Example.m

MATLAB fir2 *Function*

Another filter design function is fir2. It can be used to create filters with any arbitrary frequency response. The syntax to call this function is:

$$h = fir2(m,[0,lowF,highF,1],[dcA,lowA,highA,nyqA]);$$

- Input Variables
  - m : filter length
  - lowF, highF : normalized cutoff frequencies
  - dcA, lowA, highA, nyqA : linear amplitude at specified frequencies
- Output Variable
  - h : output impulse response

Examples: Create and run the following m-file in MATLAB.

```
% FIR2EXAMPLE
% This script demonstrates the syntax of
% the "fir2" function to create a filter with
% arbitrary frequency response
%
% See also FIR2
```

```
% Declare function parameters
n = 30;   % filter order
frqs = [0, 0.2, 0.5,0.8,1];    % normalized frequencies
amps = [2, 4, 0.25,2,1];  % linear amplitudes for each freq



% Syntax for function
h = fir2(n,frqs,amps);

% Plot frequency response
freqz(h);
```

Example 12.2: Script: fir2Example.m

12.6 *Changing the Filter Order*

The filter order, or the number of delay lines, directly influences the performance of the filter. Increasing the number of delay lines (lengthening the system's impulse response) typically leads to an improvement in characteristics such as steepness (or slope) of the transition band, flatness of the pass band, and attenuation of the stop band. In Figure 12.25, an example of the filter magnitude response is shown for the function h = fir1(m,0.75) with several different filter orders, m.

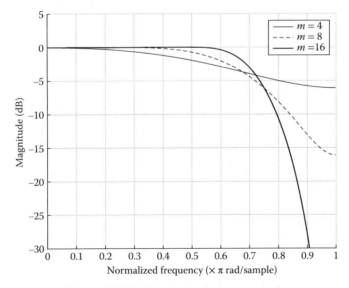

Figure 12.25: Filter order design constraint

12.7 *Processing a Signal Using an FIR Filter*

The filtering process of a signal can be performed several ways in MATLAB. One approach is to index the array of the signal based on the difference equation (Section 11.5.1). Another approach is based on the convolution operation (Section 11.8).

12.7.1 *Convolution*

If the impulse response, h, of an FIR filter is known, the convolution operation can be used with the input signal array, x, by calling the function y = conv(x,h). Therefore, convolution can be used to perform filtering, in addition to feedforward echo effects and reverberation (Section 11.8.3). In general, it is the operation used to perform the processing of any finite-length LTI system.

Examples: Create and run the following m-file in MATLAB.

```
% convolutionFiltering.m
%
% This script demonstrates how to use a built-in FIR
% filter design function to create the impulse response
% for an LPF. Then, the filtering is performed on an
% audio signal using the convolution operation.


% Import our audio file
[x,Fs] = audioread('AcGtr.wav');
Nyq = Fs/2;


n = 30; % Order of the filter


freqHz = 500; % frequency in Hz
Wn = freqHz/Nyq; % Normalized frequency for fir1


[ h ] = fir1(n,Wn);   % Filter design function


% "h" is the impulse response of the filter


% Convolution applies the filter to a signal
y = conv(x,h);


sound(y,Fs); % Listen to the effect
```

Example 12.3: Script: convolutionFiltering.m

12.8 Appendix I: Combining Multiple Filters

The following appendix includes illustrative examples of the results when multiple filters are combined together. These examples conceptually demonstrate how more complex types of spectral processors can be designed.

12.8.1 Series Filters

Using the basic, first-order LPF and HPF as building blocks, consider the following examples of combining filters together in series.

Series LPF and HPF

First, consider the block diagram shown in Figure 12.26 consisting of an LPF and HPF in series. Substituting the first-order LPF and HPF, an expanded block diagram is shown in Figure 12.27 with input $x[n]$ output $y[n]$, and an intermediary signal, $u[n]$.

The difference equation for the block diagram in Figure 12.27 can be written as an LPF: $u[n] = x[n] + x[n-1]$, and an HPF: $y[n] = u[n] - u[n-1]$. Additionally, the difference equation can be written entirely as a function of the input signal by substituting the following:

$$u[n-1] = x[n-1] + x[n-2]$$

$$y[n] = (x[n] + x[n-1]) - (x[n-1] + x[n-2])$$

$$y[n] = x[n] - x[n-2]$$

This difference equation is identical to the BPF described in Section 12.3.2. The normalized magnitude response for the separate LPF and HPF, along with the combined BPF, is shown in Figure 12.28.

Figure 12.26: Block diagram of LPF and HPF in series

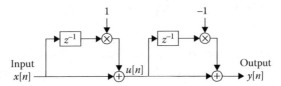

Figure 12.27: Expanded block diagram of LPF and HPF in series

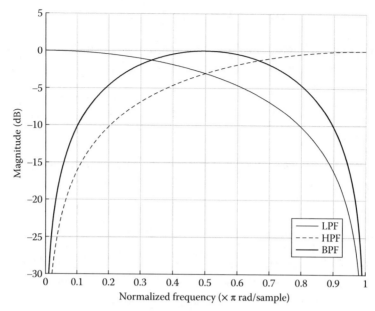

Figure 12.28: Magnitude response of LPF and HPF in series

Figure 12.29: Block diagram of two LPFs in series

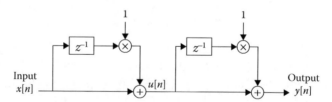

Figure 12.30: Expanded block diagram of two LPFs in series

Two Series LPFs

Next, consider the block diagram shown in Figure 12.29 consisting of two LPFs in series. By substituting the first-order LPF from Section 12.2.1, an expanded block diagram is created for Figure 12.30.

The difference equation for the block diagram in Figure 12.30 can be written as two separate LPFs: $u[n] = x[n] + x[n-1]$ and $y[n] = u[n] + u[n-1]$. Additionally, the difference equation can be written entirely as a function of the input signal by substituting the following:

$$u[n-1] = x[n-1] + x[n-2]$$

$$y[n] = (x[n] + x[n-1]) + (x[n-1] + x[n-2])$$

$$y[n] = x[n] + 2 \cdot x[n-1] + x[n-2]$$

The impulse response of the system is h = [1 2 1]. The overall system gain can be normalized resulting in the impulse response h = [0.25 0.5 0.25]. The magnitude response of the system is shown in Figure 12.31.

Compared to a single LPF, the combined system creates a low-pass filter with a different cutoff frequency and a steeper slope to the roll-off, or skirt, of the filter.

12.8.2 Parallel Filters

In addition to series filters, consider the system described in Figure 12.32 with an LPF and HPF in parallel. This system can be used to create other spectral processors, with one example being a band-stop filter (BSF). The individual filters must be designed such that the cutoff frequency of the LPF is less than the cutoff frequency of the HPF. In Example 12.4, a BSF is

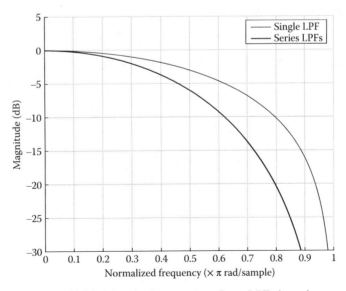

Figure 12.31: Magnitude response of two LPFs in series

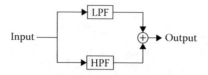

Figure 12.32: Block diagram of LPF and HPF in parallel

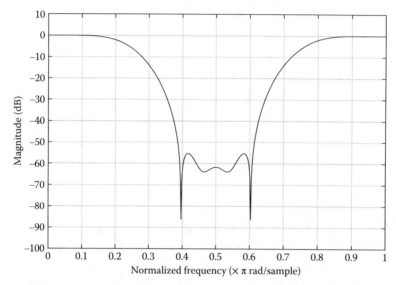

Figure 12.33: Magnitude response of BSF

created by using two different filters. The frequency response of the combined effect is shown in Figure 12.33.

Examples: Create and run the following m-file in MATLAB.

```
% BANDSTOPFILTER
% This script creates a band-stop filter by performing
% parallel processing with an LPF and HPF
%
% See also FIR1

% Design filters
% Note: W_lpf must be less than W_hpf to
% create a band-stop filter
order = 24;
W_lpf = 0.25;    % Normalized freq of LPF
lpf = fir1(order,W_lpf);
```

```
W_hpf = 0.75;    % Normalized freq of HPF
hpf = fir1(order,W_hpf,'high');

% Impulse input signal
input = [1, 0];

% Separately, find impulse response of LPF and HPF
u = conv(input,lpf);

w = conv(input,hpf);

% Create combined parallel output by adding together IRs
output = u + w;

% Plot the frequency response
freqz(output);
```

Example 12.4: Script: `bandStopFilter.m`

12.9 Appendix II: Using an FIR Filter to Synthesize Pink Noise

In Section 7.4.1, the process of synthesizing white noise is presented. White noise has an equal likelihood of amplitude at all frequencies across the spectrum. In other words, there is an equal distribution of power ($P = A^2$) between 200–400 Hz as there is between 10,200–10,400 Hz.

Another type of aperiodic signal is **pink noise**. It does not have a flat frequency spectrum when averaged over time. Rather, the amplitude of pink noise decreases with increasing frequency. Specifically, the amplitude at a given frequency, f, is $A = \frac{1}{\sqrt{f}}$, such that the signal's power for the given frequency is $P = \frac{1}{f}$ (Keshner, 1982).

Pink noise can be synthesized by filtering white noise. The spectral roll-off of pink noise is applied by the filter based on the specifications above. In Example 12.5, a filter is designed using fir2 with linear amplitude values of $\frac{1}{\sqrt{f}}$.

Examples: Create and run the following m-file in MATLAB.

```
% PINKNOISE1
% This script synthesizes an approximation of pink noise
% using an FIR filter.
%
```

```
% Pink noise can be created by filtering white noise. The
% amplitude response of the filter has a gain of 1/sqrt(f),
% where "f" is frequency (Hz).
%
% See also FIR2

clc; clear;
Fs = 48000;      % Sampling rate
Nyq = Fs/2;      % Nyquist frequency for normalization
sec = 5;         % 5 seconds of noise
white = randn(sec*Fs,1);



f = 20;      % Starting frequency in Hz
gain = 1/sqrt(f);   % Amplitude at starting frequency


freq = 0;   % Initialize fir2 freq vector
while f < Nyq
    % Normalized frequency vector
    freq = [freq f/Nyq];

    % Amplitude vector, gain = 1/sqrt(f)
    gain = [gain 1/sqrt(f)];

    % Increase "f" by an octave
    f = f*2;
end
% Set frequency and amplitude at Nyquist
freq = [freq 1];
gain = [gain 1/sqrt(Nyq)];

% Filter normalization factor to unity gain
unity = sqrt(20);
gain = unity * gain;

% Plot frequency response of filter
order = 2000;
h = fir2(order,freq,gain);
[H,F] = freqz(h,1,4096,Fs);
semilogx(F,20*log10(abs(H))); axis([20 20000 -30 0]);
```

```
% Create pink noise by filtering white noise
pink = conv(white,h);
sound(pink,Fs);
```

Example 12.5: Script: `pinkNoise1.m`

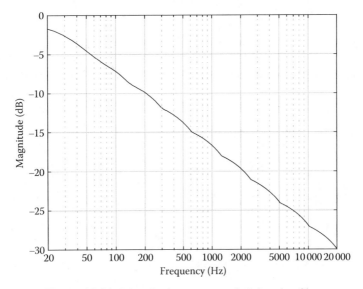

Figure 12.34: Magnitude response of pink noise filter

The spectral magnitude response of the pink noise filter is shown in Figure 12.34. This type of noise has an equal distribution of power per octave. As an example, there is an equal distribution of power between 200–400 Hz as there is between 2000–4000 Hz.

As indicated by the spectrum and shown mathematically next, pink noise has a decrease in amplitude at a rate of approximately −3 dB per octave (i.e., increase in frequency by a factor of 2), or a rate of −10 dB per decade (i.e., increase in frequency by a factor of 10).

$$20 \cdot log_{10} \left(\frac{\frac{1}{\sqrt{2 \cdot f}}}{\frac{1}{\sqrt{f}}} \right)$$

$$20 \cdot log_{10} \left(\frac{(2 \cdot f)^{-\frac{1}{2}}}{f^{-\frac{1}{2}}} \right)$$

$$20 \cdot log_{10} \left(\frac{2^{-\frac{1}{2}} \cdot f^{-\frac{1}{2}}}{f^{-\frac{1}{2}}} \right)$$

$$20 \cdot log_{10} \left(2^{-\frac{1}{2}} \right) = -3.01 \text{ dB}$$

$$\text{Similarly: } 20 \cdot log_{10} \left(\frac{\frac{1}{\sqrt{10 \cdot f}}}{\frac{1}{\sqrt{f}}} \right)$$

$$20 \cdot log_{10} \left(10^{-\frac{1}{2}} \right) = -10 \text{ dB}$$

Bibliography

Keshner, M. S. (1982). 1/*f* noise. *Proceedings of the IEEE, 70*(3), 212–218. doi:10.1109/PROC.1982.12282

Chapter 13

Infinite Impulse Response Filters
Systems with Memory: Spectral Processors (Part 2)

13.1 Introduction: Filters with Feedback

This chapter introduces the use of feedback in systems to create spectral processors. In Chapter 12, spectral processors were created using feedforward designs, exclusively. This chapter begins by examining systems that only use feedback delay. Then, feedforward and feedback designs are combined to present generalized systems with memory.

13.1.1 Spectral Analysis of Filters with Feedback

As discussed in Chapter 11, the use of feedback delay creates a system with an impulse response of infinite length. This is called an *infinite impulse response* (IIR) system. In MATLAB, an IIR system is represented by one array of the gain values for the feedback delay lines (a = $[a_0, a_1, a_2, \ldots, a_W]$), and another array of the gain values for any feedforward delay lines (b = $[b_0, b_1, b_2, \ldots, b_M]$). Then the MATLAB function freqz(b,a) can be used to analyze the spectral response of the IIR system.

13.2 Basic Feedback Filters

Several basic filters can be created using a single delay line where the system output, $y[n]$, is fed back to, and combined with, the system input, $x[n]$. In Figure 13.1, a single-sample delay block is used with feedback of unity gain.

To follow the convention in MATLAB, the difference equation for this system can be rearranged to isolate terms related to the system output, y, on the left-hand side, and terms related to the system input, x, on the right-hand side.

$$y[n] = x[n] + y[n-1]$$
$$y[n] - y[n-1] = x[n]$$

$$y[n] + (-1) \cdot y[n-1] = x[n]$$

$$a_0 \cdot y[n] + a_1 \cdot y[n-1] = b_0 \cdot x[n]$$

The coefficients of the difference equation can be placed in arrays, a and b, for further analysis.

$$a = [a_0, a_1] = [1, -1]$$

$$b = [b_0] = [1]$$

The frequency response of the system in Figure 13.1 can be plotted using the command `freqz(1,[1,-1])`. The magnitude and phase response are shown in Figure 13.2. In this case, the positive feedback causes a peak in the spectral response for low frequencies.

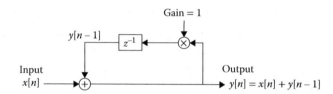

Figure 13.1: Block diagram of feedback delay with unity gain

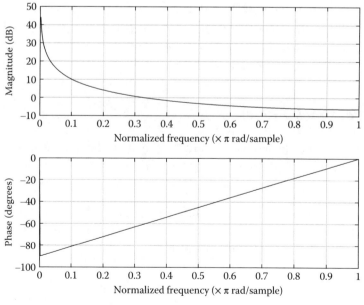

Figure 13.2: Frequency response of feedback delay with unity gain

13.2.1 Negative Feedback

One modification of this system is to invert the gain in the feedback path. This is called **negative feedback**. A block diagram for this system is shown in Figure 13.3.

The following describes the difference equation for this system and the coefficients to be used to define this system as arrays:

$$y[n] = x[n] - y[n-1]$$
$$y[n] + y[n-1] = x[n]$$
$$a_0 \cdot y[n] + a_1 \cdot y[n-1] = b_0 \cdot x[n]$$
$$\mathsf{a} = [a_0, a_1] = [1, 1]$$
$$\mathsf{b} = [b_0] = [1]$$

The frequency response of the system in Figure 13.3 can be plotted using the command `freqz(1,[1,1])`. The magnitude and phase response are shown in Figure 13.4. In this case, the negative feedback causes a peak in the spectral response for high frequencies.

To further explore the performance of this system, consider if the gain in the feedback path can take on values ranging from $\alpha = -1$ to 0. This system and difference equation is shown in Figure 13.5.

The frequency response of the system in Figure 13.5 can be plotted using the command `freqz(1,[1,α])`. The magnitude and phase responses for various values of α are shown in Figure 13.6. Note that the peak in the magnitude response is decreased as α approaches 0.

13.2.2 Inverted Comb Filter

Another modification to the system with feedback delay is to change the delay time in the feedback path. Rather than using one sample of delay, the delay time could be an arbitrary length, W. The block diagram for this system and the difference equation is shown in Figure 13.7.

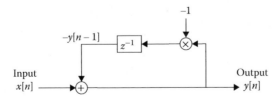

Figure 13.3: Block diagram of feedback delay with inverted gain

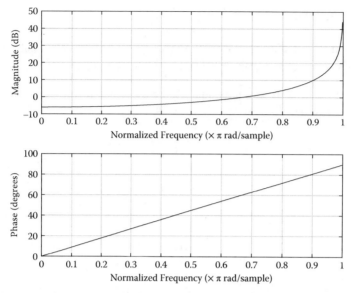

Figure 13.4: Frequency response of feedback delay with inverted gain

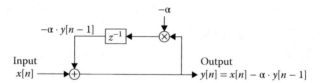

Figure 13.5: Block diagram of feedback delay with an arbitrary gain value

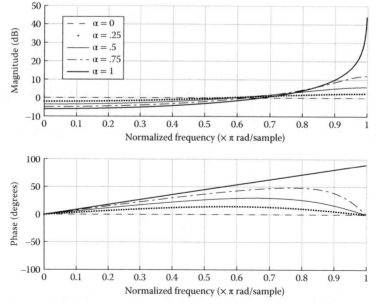

Figure 13.6: Frequency response of feedback delay for various gain values

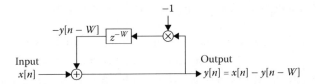

Figure 13.7: Block diagram of feed-back delay with arbitrary delay time

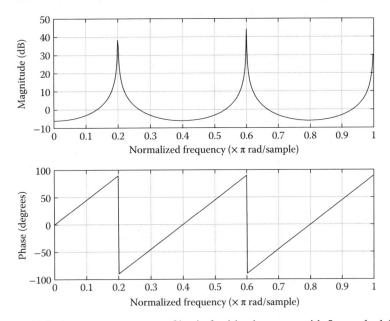

Figure 13.8: Frequency response of basic feed-back system with 5-sample delay

This system is called an inverted comb filter due to the shape of the frequency response. Specifically, the shape is the inverse of the feedforward comb filter shown in Figure 12.9.

As an example, consider the feedback system with a delay time of $W = 5$ samples. The frequency response of the system in Figure 13.7 can be plotted using the command `freqz(1,[1,0,0,0,0,1])`. The magnitude and phase response are shown in Figure 13.8.

The following relationship can be used to determine the frequencies at which peaks occur in the frequency spectrum. For the system in Figure 13.7 with negative feedback and W samples of delay, a peak occurs at frequency f relative to the Nyquist frequency for odd whole numbers, $w \leq W$:

$$f = \frac{w}{W} * Nyquist \tag{13.1}$$

For the same system with positive feedback, peaks occur at frequencies for even whole numbers, $w \leq W$.

13.3 Combined Feedforward and Feedback Filters

Thus far in Chapter 13, basic systems have been explored exclusively using feedback delay. This follows from Chapter 12, which examined systems exclusively using feedforward delay. The next step is to analyze systems with both feedforward and feedback delay.

It is worth noting that these systems are still considered IIR systems because the impulse response of the entire system is not finite. Therefore, the methods used previously to analyze IIR systems remain appropriate.

The difference equation for an arbitrary system with feedforward and feedback delay can be written:

$$y[n] = b_0 \cdot x[n] + b_1 \cdot x[n-1] + \ldots + b_M \cdot x[n-M] - a_1 \cdot y[n-1] - \ldots - a_W \cdot y[n-W]$$

$$y[n] + a_1 \cdot y[n-1] + \ldots + a_W \cdot y[n-W] = b_0 \cdot x[n] + b_1 \cdot x[n-1] + \ldots + b_M \cdot x[n-M]$$

As a convention, the coefficient for the system output, $a_0 \cdot y[n]$, is normalized to be $a_0 = 1$. Therefore, a system with feedforward and feedback delay can be represented in MATLAB using the following arrays:

$$\mathtt{b} = [b_0, b_1, \ldots, b_M]$$

$$\mathtt{a} = [1, a_1, \ldots, a_W]$$

For these systems, it is not necessary for the maximum delay time, M, of the feedforward delay lines to be equal to the maximum delay time, W, of the feedback delay lines.

13.3.1 All-Pass Filter

As an initial case, consider a system with a one-sample feedforward delay line and a one-sample feedback delay line. This system is used as a building block for more complex types of spectral processors. By intentionally selecting the gain coefficients for each branch of the system, an all-pass filter (APF) can be created. Modifications of the gain coefficients allow for the intuitive design of other types of filters.

An APF is a system that does not change the relative amplitude of frequencies contained in an input signal. However, an APF typically introduces a frequency-dependent phase shift to the input signal. The block diagram for one example of an APF is shown in Figure 13.9.

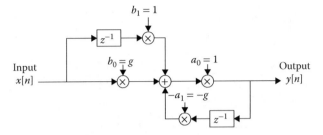

Figure 13.9: Block diagram of APF

The difference equation for the system in Figure 13.9 and related MATLAB arrays are:

$$y[n] = b_0 \cdot x[n] + b_1 \cdot x[n-1] - a_1 \cdot y[n-1]$$

$$y[n] + g \cdot y[n-1] = g \cdot x[n] + x[n-1]$$

$$b = [g, 1]$$

$$a = [1, g]$$

To create nontrivial results, the gain coefficient should have values $0 < |g| < 1$. Note the arrays, a and b, have identical values in reverse order. This pattern can also be used to design APFs for systems with an additional number of delay lines.

Frequency Response of APF

To understand the frequency response of the all-pass filter, consider the behavior of the feedforward delay and feedback delay individually. The magnitude and phase response for each part of the APF is shown separately in Figure 13.10. The feedforward delay provides an LPF magnitude response. The feedback delay provides a magnitude response exactly inverted to the LPF.

By using the LPF and feedback delay together, the result is an APF with the frequency response shown in Figure 13.11. The combined magnitude response results in a flat amplitude across the spectrum. However, the combined phase response indicates a frequency-dependent phase shift.

A similar result can found by using a feedforward HPF, b = [1 -g], with its spectral inverse feedback filter, a = [-g 1]. Separate frequency response curves are shown in Figure 13.12. The combined frequency response indicates an APF with a phase shift at low frequencies, shown in Figure 13.13.

There are many different APFs that can be created by modifying the delay length and gains of the system. Each variation can have a different phase response, but all have a flat magnitude response. Additional topologies for APFs are presented in Appendix 13.7.1.

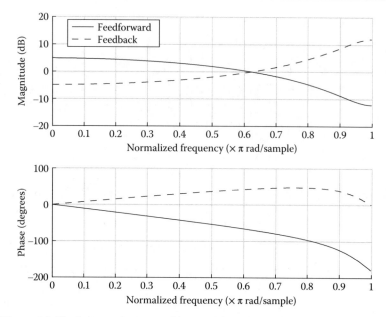

Figure 13.10: Separate low-pass filter and feed-back delay to create an APF

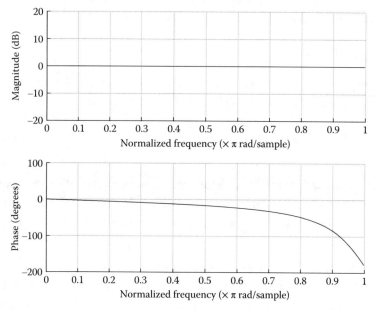

Figure 13.11: Frequency response of APF with high-frequency phase shift

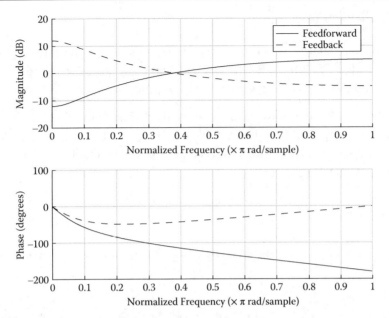

Figure 13.12: Separate high-pass filter and feed-back delay to create an APF

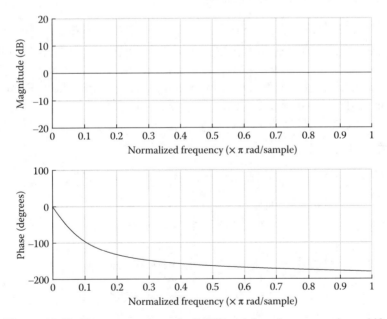

Figure 13.13: Frequency response of APF with low-frequency phase shift

13.3.2 Bi-Quadratic Filter

An important filter for audio engineering applications is the **bi-quadratic (bi-quad) filter**. It can be used to create LPFs, HPFs, BPFs, APFs, and notch filters, as well as peaking filters and shelving filters commonly found on audio mixing console equalizers (Section 13.5.1).

The bi-quad filter has two feedforward and two feedback delay lines. In both cases, one delay line introduces one sample of delay and the other delay line introduces two samples of delay. There are independent gain coefficients for each path of the filter. A block diagram for the bi-quad filter is shown in Figure 13.14.

The difference equation for the bi-quad filter is the following:

$$y[n] = b_0 \cdot x[n] + b_1 \cdot x[n-1] + b_2 \cdot x[n-2] - a_1 \cdot y[n-1] - a_2 \cdot y[n-2]$$

The gain coefficients of the bi-quad filter can be calculated based on the parameters found on audio equalizers: frequency, Q, and gain (dB). Each type of filter has a unique calculation of the gain coefficients (Bristow-Johnson, 2004).

First, the desired frequency, f, in Hz is converted to a normalized frequency, $w_0 = \frac{2 \cdot pi \cdot f}{F_s}$. Next, the Q is converted to a normalized bandwidth, $\alpha = \frac{sin(w_0)}{2 \cdot Q}$. Then, the decibel gain should be converted to a linear gain, $A = \sqrt{10^{\left(\frac{gain_{dB}}{20}\right)}}$, for peaking and shelving filters. Finally, the coefficients can be determined based on Table 13.1.

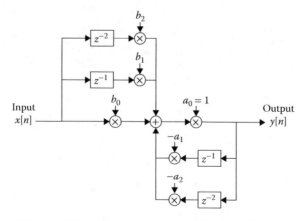

Figure 13.14: Block diagram of a bi-quadratic filter

Table 13.1: Calculation of bi-quad coefficients

	LPF	HPF	Low Shelf
b_0	$\frac{1-cos(w_0)}{2}$	$\frac{1+cos(w_0)}{2}$	$A \cdot ((A+1)-(A-1)\cdot cos(w_0)+2\sqrt{A}\cdot\alpha)$
b_1	$1-cos(w_0)$	$-1+cos(w_0)$	$2\cdot A((A-1)-(A+1)\cdot cos(w_0))$
b_2	$\frac{1-cos(w_0)}{2}$	$\frac{1+cos(w_0)}{2}$	$A\cdot((A+1)-(A-1)\cdot cos(w_0)-2\sqrt{A}\cdot\alpha)$
a_0	$1+\alpha$	$1+\alpha$	$(A+1)+(A-1)\cdot cos(w_0)+2\sqrt{A}\cdot\alpha$
a_1	$-2\cdot cos(w_0)$	$-2\cdot cos(w_0)$	$-2\cdot((A-1)+(A+1)\cdot cos(w_0))$
a_2	$1-\alpha$	$1-\alpha$	$(A+1)+(A-1)\cdot cos(w_0)-2\sqrt{A}\cdot\alpha$

	BPF1	BPF2	High Shelf
b_0	$Q\cdot\alpha$	α	$A\cdot((A+1)-(A-1)\cdot cos(w_0)+2\sqrt{A}\cdot\alpha)$
b_1	0	0	$-2\cdot A((A-1)-(A+1)\cdot cos(w_0))$
b_2	$-Q\cdot\alpha$	$-\alpha$	$A\cdot((A+1)-(A-1)\cdot cos(w_0)-2\sqrt{A}\cdot\alpha)$
a_0	$1+\alpha$	α	$(A+1)-(A-1)\cdot cos(w_0)+2\sqrt{A}\cdot\alpha$
a_1	$-2\cdot cos(w_0)$	$-2\cdot cos(w_0)$	$2\cdot((A-1)-(A+1)\cdot cos(w_0))$
a_2	$1-\alpha$	$1-\alpha$	$(A+1)-(A-1)\cdot cos(w_0)-2\sqrt{A}\cdot\alpha$

	APF	Peaking/Bell	Notch
b_0	$1-\alpha$	$1+\alpha\cdot A$	1
b_1	$-2\cdot cos(w_0)$	$-2\cdot cos(w_0)$	$-2\cdot cos(w_0)$
b_2	$1+\alpha$	$1-\alpha\cdot A$	1
a_0	$1+\alpha$	$1+\frac{\alpha}{A}$	$1+\alpha$
a_1	$-2\cdot cos(w_0)$	$-2\cdot cos(w_0)$	$-2\cdot cos(w_0)$
a_2	$1-\alpha$	$1-\frac{\alpha}{A}$	$1-\alpha$

Additional approaches for implementing a bi-quad filter are presented in Appendix 13.7.3, along with an example MATLAB function.

13.4 MATLAB IIR Filter Design Functions

There are several built-in MATLAB functions for designing IIR filters. These functions return arrays, b and a, which represent the feedforward and feedback gain coefficients.

13.4.1 MATLAB butter Function

The **Butterworth filter** design function, butter, can create LPFs, HPFs, BPFs, and band-stop filters. Butterworth filters are designed to have a flat pass band. The syntax to call this function is:

- [b,a] = butter(m,Wn,ftype);
  - m : filter length
  - Wn : normalized cutoff frequency

- ftype: filter type—'low', 'high', 'pass', 'stop'
- b : numerator of transfer function
- a : denominator of transfer function

13.4.2 MATLAB `ellip` *Function*

The elliptic filter design function, `ellip`, can also be used to create LPFs, HPFs, BPFs, and band-stop filters. Elliptic filters do not necessarily have a flat pass band. Rather, coefficients are determined to create a filter with the steepest possible slope, given a filter order, tolerance of ripple in the pass band, and desired attenuation in the stop band. The syntax to call this function is:

- `[b,a] = ellip(m,Rp,Rs,Wp,ftype);`
  - m : filter length
  - Rp : allowable pass-band ripple in dB
  - Rs : stop-band attenuation
  - Wp : normalized cutoff frequency [0,1]
  - ftype: filter type—'low', 'high', 'pass', 'stop'
  - b : numerator of transfer function
  - a : denominator of transfer function

13.4.3 *Filter Order, Ripple, Slope, Attenuation*

There are several inherent constraints, or trade-offs, for filter design. Generally speaking, the response of a filter with a higher order can have a better fit to a set of desired features than a filter with a lower order (Section 12.6). However, there is a computational cost to processing a signal with a higher filter order. One benefit of IIR filters compared to FIR filters is IIR filters with a relatively low order can achieve a response only matched by FIR filters with a much higher order.

Besides the filter order, the various features of the filter also introduce the possibility to trade-off performance. The ripple in the pass band, the slope in the transition band, and the attenuation in the stop band all act as degrees of freedom in designing the optimum filter. In some cases, an audio engineer may be willing to sacrifice the filter slope to achieve a flatter pass band. In other cases, it may be preferable to have a steeper slope at the cutoff frequency, but sacrifice the stop-band attenuation.

An example is shown in Figure 13.15 of the filter design specifications. The LPF was designed to meet the specifications of the allowable pass-band ripple, R_p, the minimum amount of attenuation in the stop band, R_s, and the slope of the transition band. The slope is determined by the cutoff frequency of the pass band, f_{ps}, as well as the frequency of the stop band, f_{st}, where the minimum attenuation requirement is initially met.

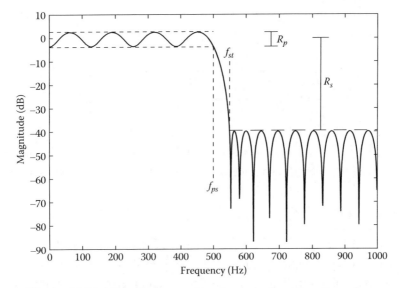

Figure 13.15: Filter design specifications: ripple, attenuation, slope

Figure 13.16: Block diagram of an audio equalizer effect

13.5 Series and Parallel Filter Configurations

Individual filters can be combined together to create more complex spectral effects. The audio equalizer is based on a series combination of filters, while multi-band processing is accomplished using parallel filters.

13.5.1 Equalizer

An equalizer is an audio effect created by cascading multiple filters in series. In the equalizer, each filter is called a **band**. A channel strip on many audio mixing consoles includes a 3-band, 4-band, or 5-band equalizer. Each band can be a different type of filter. An example of a block diagram to create an audio equalizer is shown in Figure 13.16. Due to its flexibility, a bi-quad filter can be used for each individual band of the equalizer.

Graphic Equalizer

A **graphic equalizer** is a specific type of audio equalizer. In this effect, multiple peaking/bell filters are combined in series. Each filter has a variable gain control, but a fixed frequency and Q. Therefore, the graphic equalizer can be contrasted with the **parametric equalizer**, which

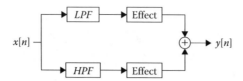

Figure 13.17: Block diagram of a basic multi-band effect

has variable control for all three parameters. Audio engineers use graphic equalizers in live sound reinforcement to spectrally balance the output of the mix for the system and room.

13.5.2 Multi-Band Processing

Multi-band processing is a general approach to using an audio effect in distinct frequency regions. There are many examples used by audio engineers including multi-band distortion, multi-band compression, multi-band stereo widening, and the vocoder (multi-band envelope follower, Section 17.5).

For all of these effects, an input signal is sent through a group of parallel filters called a **filter bank**. The cutoff frequencies of the filters are selected such that the resulting frequency bands are contiguous. After the signal is filtered, additional processing can be performed on each frequency band. As an example, this makes it possible to apply one distortion to low frequencies and a different distortion to high frequencies in the case of multi-band distortion. Following the separate processing, all the frequency bands can be added back together to create the output signal. In Figure 13.17, a block diagram of an arbitrary multi-band processor is shown with two filters. An implementation of a two-band filter bank is demonstrated in Example 13.1. The frequency response of each filter is shown in Figure 13.18.

Examples: Create and run the following m-file in MATLAB.

```
% BASICFILTERBANK
% This script creates a two-band filter bank using
% an LPF and HPF. The Butterworth filter design
% function is used to create the LPF and HPF,
% both with the same cutoff frequency.
% The magnitude response of each filter is
% plotted together.
%
% See also BUTTER, FREQZ

Fs = 48000;
Nyq = Fs/2;
```

```
n = 8;
Wn = 1000/Nyq;


[bLow,aLow] = butter(n,Wn);
[bHi,aHi] = butter(n,Wn,'high');


[hLow,w] = freqz(bLow,aLow,4096,Fs);
[hHi] = freqz(bHi,aHi,4096,Fs);


semilogx(w,20*log10(abs(hLow)),w,20*log10(abs(hHi)));
axis([20 20000 -24 6]);
xlabel('Frequency (Hz)');
ylabel('Amplitude (dB)');
legend('LPF','HPF');
```

Example 13.1: Script: `basicFilterbank.m`

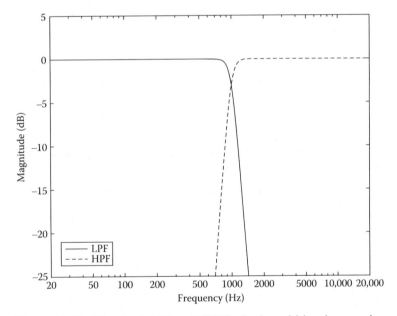

Figure 13.18: Filter bank LPF and HPF for basic multi-band processing

The routing of a mutli-band processor can be generalized for *B* number of filters and frequency bands. A block diagram for a filter bank with *B* bands is shown in Figure 13.19. A MATLAB implementation is demonstrated in Example 13.2 for creating a filter bank with logarithmically spaced cutoff frequencies. The magnitude response is shown in Figure 13.20.

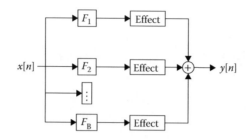

Figure 13.19: Block diagram of a generalized multi-band effect

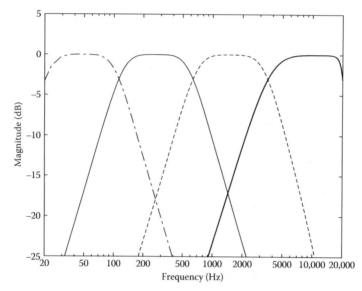

Figure 13.20: Magnitude response of logarithmic filter bank for multi-band processing

Examples: Create and run the following m-file in MATLAB.

```
% FILTERBANKEXAMPLE
% This script creates a two-band filter bank using
% an LPF and HPF
%
% See also BUTTER

clear; clc;

Fs = 48000;
Nyq = Fs/2;
m = 2; % Filter order

numOfBands = 4;
```

```
% Logarithmically spaced cutoff frequencies
% 2*10^1 - 2*10^4 (20-20k) Hz
freq = 2 * logspace(1,4,numOfBands+1);


for band = 1:numOfBands


    Wn = [freq(band) , freq(band+1)] ./ Nyq;
    [b(:,band),a(:,band)] = butter(m,Wn);


    [h,w] = freqz(b(:,band),a(:,band),4096,Fs);
    semilogx(w,20*log10(abs(h)));
    hold on;


end


hold off;
axis([20 20000 -24 6]);
xlabel('Frequency (Hz)');
ylabel('Amplitude (dB)');
```

Example 13.2: Script: filterbankExample.m

13.6 *Processing a Signal Using an IIR Filter*

Because of the feedback in an IIR filter, it is not possible to use the convolution operation with digital signals. In MATLAB, there is a separate function, filter, which can be used to process a signal. The feedfoward and feedback gain coefficients are included as input arguments to the function using the following syntax: y = filter(b,a,x).

Examples: Create and run the following m-file in MATLAB.

```
% FILTEREXAMPLE
%
% This script demonstrates how to use a built-in IIR
% filter design function to create the impulse response
% for an LPF. Then, the filtering is performed on an
% audio signal using the "filter" function.
%
% See also BUTTER, FILTER


% Import our audio file
```

```
[x,Fs] = audioread('AcGtr_1.wav');
Nyq = Fs/2;


m = 4; % Order of the filter


freqHz = 500; % frequency in Hz
Wn = freqHz/Nyq;


[b,a] = butter(m,Wn);


y = filter(b,a,x);


sound(y,Fs); % Listen to the effect
```

Example 13.3: Script: filterExample.m

13.6.1 Approximating an IIR Filter as an FIR Filter

An alternative approach to filtering a signal with an IIR system is to approximate the IIR system as an FIR system. The length of the finite impulse response is a truncated version of the IIR system based on an impulse response measurement. Ideally, the length of the FIR should be selected to represent the filter delay taps as long as their amplitude is greater than a minimum threshold.

The MATLAB function in the Signal Processing Toolbox to measure the impulse response of an IIR system has the syntax [h] = impz(b,a). After the FIR system, h, is approximated, the convolution operation can be used to perform filtering of a signal.

Examples: Create and run the following m-file in MATLAB.

```
% IMPZEXAMPLE
% This script demonstrates how to use the built-in function
% "impz" to approximate an IIR system as an FIR system.
% Then filtering is performed on an audio
% signal using the convolution operation.
%
% See also IMPZ, CONV, BUTTER

% Import our audio file
[x,Fs] = audioread('AcGtr_1.wav');
Nyq = Fs/2;
```

```
m = 4; % Order of the filter


freqHz = 2000; % frequency in Hz
Wn = freqHz/Nyq;


[b,a] = butter(m,Wn);


h = impz(b,a); % Approximate system


y = conv(x,h);


sound(y,Fs); % Listen to the effect


stem(h);
```

Example 13.4: Script: `impzExample.m`

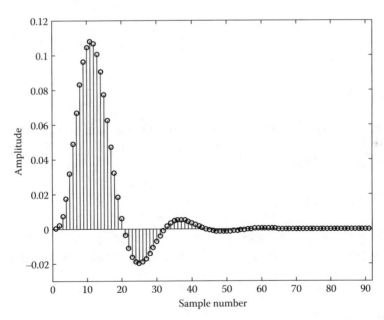

Figure 13.21: Approximation of an IIR system as a FIR sytem using the `impz` function

In Figure 13.21, an approximation of an IIR low-pass filter is shown as an FIR system. For this IIR filter with an order of 4, the equivalent FIR filter has an approximate order of 100.

13.7 Appendix I: Additional Filter Forms

Filters with both feedforward and feedback delay can by implemented in multiple forms, or topologies. Previous examples in this chapter used separate blocks for each delay path, which allowed the filter's difference equation to be written directly from the block diagram. However, this form is not ideal for practical implementation. Other approaches produce identical processing with fewer delay blocks for improved efficiency.

13.7.1 APF Direct Form II

One form of an all-pass filter is presented in Section 13.3.1. The **Direct Form II** APF rearranges the routing of the input signal such that only one delay block is necessary, saving memory and computation when implemented. Despite the changes to the block diagram (shown in Figure 13.22), the difference equation for the output, $y[n]$, is identical to the APF with separate feedforward and feedback delay blocks.

$$y[n] = g \cdot x[n] + w[n-1]$$
$$w[n] = (-g) \cdot y[n] + x[n]$$
$$y[n] = g \cdot x[n] + (-g) \cdot y[n-1] + x[n-1]$$

This form of the APF is used for phaser effects (Section 15.7.1) and algorithmic reverb effects (Section 16.2.2).

13.7.2 Nested APF

Another approach to create an APF involves nesting a Direct Form II APF within the topology of another APF (Figure 13.23). This is an alternative method to using series or parallel routing to connect two filters together.

13.7.3 Bi-Quad Filter Topologies

Due to the ubiquity of the bi-quad filter for audio applications, it is worth noting several of the various forms in which it can be implemented. In Example 13.6, each of the following

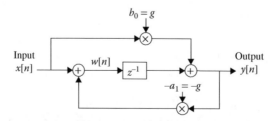

Figure 13.22: Block diagram of Direct Form II APF

three forms of the bi-quad filter is included in a MATLAB function for comparison purposes.

Direct Form I

In Figure 13.24, the **Direct Form I** topology is shown. It is similar in layout to the form in Section 13.3.2, with the differences being the delay blocks are all one sample in duration and cascade in series to accomplish two combined samples of delay. This approach reduces the necessary number of samples stored in memory.

This form leads to the following update equations to calculate the output, $y[n]$, for the current sample, n, within a loop. In this topology, it is necessary to store four samples, χ_1, χ_2, Υ_1, and Υ_2 for the next loop iteration.

$$y[n] = b_0 \cdot x[n] + b_1 \cdot \chi_1 + b_2 \cdot \chi_2 + (-a_1) \cdot \Upsilon_1 + (-a_2) \cdot \Upsilon_2$$

$$\chi_2 = \chi_1$$

$$\chi_1 = x[n]$$

$$\Upsilon_2 = \Upsilon_1$$

$$\Upsilon_1 = y[n]$$

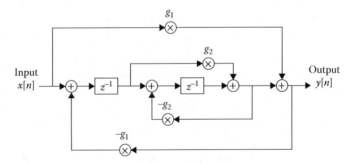

Figure 13.23: Block diagram of Direct Form II nested APF

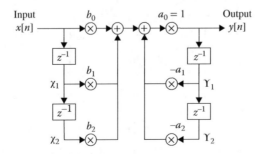

Figure 13.24: Block diagram of Direct Form I bi-quadratic filter

Direct Form II

In Figure 13.25, the **Direct Form II** topology is shown using only two delay blocks, each with one sample of delay. By routing the output of one delay block into the other delay block, a minimum amount of memory is used for the necessary delay time.

The following update equations can be used within a loop to calculate the output, $y[n]$, for the current sample, n. In this topology, it is only necessary to store two samples, $w[n-1]$ and $w[n-2]$, for the next loop iteration.

$$w[n] = x[n] + (-a_1) \cdot w[n-1] + (-a_2) \cdot w[n-2]$$

$$y[n] = b_0 \cdot w[n] + b_1 \cdot w[n-1] + b_2 \cdot w[n-2]$$

$$w[n-2] = w[n-1]$$

$$w[n-1] = w[n]$$

Transposed Direct Form II

Another variation of the bi-quad filter is called **Transposed Direct Form II**. This topology is shown in Figure 13.26. The output of the delay blocks are labeled γ_1 and γ_2, respectively.

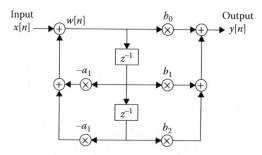

Figure 13.25: Block diagram of Direct Form II bi-quadratic filter

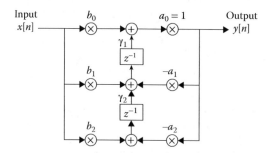

Figure 13.26: Block diagram of Transposed Direct Form II bi-quadratic filter

This form leads to the following update equations to calculate the output, $y[n]$, for the current sample, n, within a loop. In this topology, it is only necessary to store two samples, γ_1 and γ_2, for the next loop iteration.

$$y[n] = b_0 \cdot x[n] + \gamma_1$$

$$\gamma_1 = b_1 \cdot x[n] + (-a_1) \cdot y[n] + \gamma_2$$

$$\gamma_2 = b_2 \cdot x[n] + (-a_2) \cdot y[n]$$

Bi-Quad MATLAB Function Implementation

The bi-quad filter is implemented as a MATLAB function in Example 13.6. The function calculates coefficients for the filter types based on Table 13.1. The topology can be chosen from Direct Form I, Direct Form II, and Transposed Direct Form II implementations to process an input signal. A test script is used in Example 13.5 to demonstrate the bi-quad function.

Examples: Create and run the following m-files in MATLAB.

```
% BIQUADEXAMPLE
% This script demonstrates the use of the
% bi-quad filter function. Various filter
% types and topologies can be tested.
%
% See also BIQUADFILTER

% Impulse response of bi-quad
x = [1;zeros(4095,1)];

% Filter parameters
Fs = 48000;
f = 1000;       % Frequency in Hz
Q = 0.707;
dBGain = -6;

% FILTER TYPE >>> lpf,hpf,pkf,apf,nch,hsf,lsf,bp1,bp2
type = 'lpf';
% TOPOLOGY >>> 1 = Direct Form I, 2 = II, 3 = Transposed II
form = 3;


y = biquadFilter(x,Fs,f,Q,dBGain,type,form);

% Plot amplitude response of filter
```

```
[h,w] = freqz(y,1,4096,Fs);
semilogx(w,20*log10(abs(h)));axis([20 20000 -20 15]);
xlabel('Frequency (Hz)'); ylabel('Amplitude (dB)');
```

Example 13.5: Script: biquadExample.m

```
% BIQUADFILTER
% This function implements a bi-quad filter based
% on the Audio EQ Cookbook Coefficients. All filter
% types can be specified (LPF, HPF, BPF, etc.) and
% three different topologies are included.
%
% Input Variables
%   f0 : filter frequency (cutoff or center based on filter)
%   Q : bandwidth parameter
%   dBGain : gain value on the decibel scale
%   type : 'lpf','hpf','pkf','bp1','bp2','apf','lsf','hsf'
%   form : 1 (Direct Form I), 2 (DFII), 3 (Transposed DFII)

function [out] = biquadFilter(in,Fs,f0,Q,dBGain,type,form)

%%% Initial parameters
N = length(in);
out = zeros(length(in),1);

%%% Intermediate variables
%
w0 = 2*pi*f0/Fs;          % Angular freq. (radians/sample)
alpha = sin(w0)/(2*Q);    % Filter width
A  = sqrt(10^(dBGain/20)); % Amplitude

%%%%%%%%%%%%%%%%%%%%%%%%%%%%%%%%%%%%%%%%%%%%%
%%% TYPE - LPF,HPF,BPF,APF,HSF,LSF,PKF,NCH
%
%----------------------
%         LPF
%----------------------
if strcmp(type,'lpf')
    b0 =  (1 - cos(w0))/2;
```

```
    b1 =   1 - cos(w0);
    b2 =  (1 - cos(w0))/2;
    a0 =   1 + alpha;
    a1 =  -2*cos(w0);
    a2 =   1 - alpha;


%----------------------
%         HPF
%----------------------
elseif strcmp(type,'hpf')
    b0 =  (1 + cos(w0))/2;
    b1 = -(1 + cos(w0));
    b2 =  (1 + cos(w0))/2;
    a0 =   1 + alpha;
    a1 =  -2*cos(w0);
    a2 =   1 - alpha;


%--------------------------------
%   Peaking Filter
%----------------------
elseif strcmp(type,'pkf')
    b0 =   1 + alpha*A;
    b1 =  -2*cos(w0);
    b2 =   1 - alpha*A;
    a0 =   1 + alpha/A;
    a1 =  -2*cos(w0);
    a2 =   1 - alpha/A;


%----------------------
%   Band-Pass Filter 1
%----------------------
% Constant skirt gain, peak gain = 0
elseif strcmp(type,'bp1')
    b0 =   sin(w0)/2;
    b1 =   0;
    b2 =  -sin(w0)/2;
    a0 =   1 + alpha;
    a1 =  -2*cos(w0);
    a2 =   1 - alpha;
```

```
%----------------------
%    Band-Pass Filter 2
%----------------------
% Constant 0 dB peak gain
elseif strcmp(type,'bp2')
    b0 =    alpha;
    b1 =    0;
    b2 =   -alpha;
    a0 =    1 + alpha;
    a1 =   -2*cos(w0);
    a2 =    1 - alpha;


%----------------------
%    Notch Filter
%----------------------
elseif strcmp(type,'nch')
    b0 =    1;
    b1 =   -2*cos(w0);
    b2 =    1;
    a0 =    1 + alpha;
    a1 =   -2*cos(w0);
    a2 =    1 - alpha;


%----------------------
%    All-Pass Filter
%----------------------
elseif strcmp(type,'apf')
    b0 =    1 - alpha;
    b1 =   -2*cos(w0);
    b2 =    1 + alpha;
    a0 =    1 + alpha;
    a1 =   -2*cos(w0);
    a2 =    1 - alpha;


%----------------------
%    Low-Shelf Filter
%----------------------
elseif strcmp(type,'lsf')
    b0 = A*((A+1) - (A-1)*cos(w0) + 2*sqrt(A)*alpha);
    b1 = 2*A*((A-1) - (A+1)*cos(w0));
```

```
        b2 = A*((A+1) - (A-1)*cos(w0) - 2*sqrt(A)*alpha);
        a0 = (A+1) + (A-1)*cos(w0) + 2*sqrt(A)*alpha;
        a1 = -2*((A-1) + (A+1)*cos(w0));
        a2 = (A+1) + (A-1)*cos(w0) - 2*sqrt(A)*alpha;


%--------------------------------
%    High-Shelf Filter
%--------------------------
elseif strcmp(type,'hsf')
        b0 = A*( (A+1) + (A-1)*cos(w0) + 2*sqrt(A)*alpha);
        b1 = -2*A*((A-1) + (A+1)*cos(w0));
        b2 = A*((A+1) + (A-1)*cos(w0) - 2*sqrt(A)*alpha);
        a0 = (A+1) - (A-1)*cos(w0) + 2*sqrt(A)*alpha;
        a1 = 2*((A-1) - (A+1)*cos(w0));
        a2 = (A+1) - (A-1)*cos(w0) - 2*sqrt(A)*alpha;


% Otherwise, no filter
else
        b0 = 1; a0 = 1;
        b1 = 0; b2 = 0; a1 = 0; a2 = 0;
end


%%%%%%%%%%%%%%%%%%%%%%%%%%%%%%%%%%%%%%%%%%%%%%%%%%%%%
%%% Topology - Direct Form I, II, Transposed II
if (form == 1) % Direct Form I
    x2 = 0;     % Initial conditions
    x1 = 0;
    y2 = 0;
    y1 = 0;
    for n = 1:N
        out(n,1) = (b0/a0)*in(n,1) + (b1/a0)*x1 + (b2/a0)*x2 ...
            + (-a1/a0)*y1 + (-a2/a0)*y2;
        x2 = x1;
        x1 = in(n,1);
        y2 = y1;
        y1 = out(n,1);
    end


elseif (form == 2) % Direct Form II
    w1 = 0;       % w1 & w2 are delayed versions of "w"
```

```
    w2 = 0;
    for n = 1:N
        w = in(n,1) + (-a1/a0)*w1 + (-a2/a0)*w2;
        out(n,1) = (b0/a0)*w + (b1/a0)*w1 + (b2/a0)*w2;
        w2 = w1;
        w1 = w;
    end

elseif (form == 3) % Transposed Direct Form II
    d1 = 0;    % d1 & d2 are outputs of the delay blocks
    d2 = 0;
    for n = 1:N
        out(n,1) = (b0/a0)*in(n,1) + d1;
        d1 = (b1/a0)*in(n,1) + (-a1/a0)*out(n,1) + d2;
        d2 = (b2/a0)*in(n,1) + (-a2/a0)*out(n,1);
    end

else % No filtering

    out = in;

end
```

Example 13.6: Function: `biquadFilter.m`

13.8 Appendix II: Slew Rate Distortion

The use of feedback makes possible the implementation of a type of distortion called **slew rate distortion**. It is a characteristic of operational amplifiers (op-amps) used in audio electronics, and can be modeled in computer code. An op-amp has an intrinsic limitation to the rate at which amplitude can change. This characteristic is called the slew rate.

For signals where the rate at which the amplitude changes is greater than the slew rate, an op-amp cannot process the signal fast enough causing a distortion. However, if the rate at which amplitude changes is less than the slew rate, the signal is not distorted.

All else being equal, high-frequency signals change amplitude at a faster rate than low-frequency signals. Therefore, the slew rate, SR, of an op-amp is directly related to the maximum frequency, f_{max}, it can process without distortion.

$$\frac{SR \ radians}{sec} = \frac{f_{max} \ cycles}{sec} \cdot \frac{2\pi \ radians}{cycle}$$

When implemented for digital signals, the slew rate can be converted to a maximum amount of amplitude change per sample. In other words, the slew rate sets the maximum slope, s_Δ, between samples.

$$\frac{s_\Delta \; radians}{sample} = \frac{SR \; radians}{sec} \cdot \frac{T_s \; sec}{sample}$$

The slope between the current input sample, $x[n]$, and the previous output sample, $y[n-1]$, is given by $x[n] - y[n-1]$. If the slope is less than s_Δ, then the output of the current sample is: $y[n] = x[n]$. However, if the slope is greater than s_Δ, then the output of the current sample is: $y[n] = y[n-1] + s_\Delta$, which is the slew rate distortion.

Examples: Create and run the following m-files in MATLAB.

```
% SLEWRATEEXAMPLE
% This script demonstrates how to use the
% slewRateDistortion function. Two examples are
% provided, with a sine wave and a square wave.
%
% See also SLEWRATEDISTORTION

% Initial parameters
Fs = 48000; Ts = 1/Fs;
f = 5;    % Low frequency for sake of plotting
t = [0:Ts:1].';

x = sin(2*pi*f*t);    % Example 1: Sine Wave
%x = square(2*pi*f*t); % Example 2: Square Wave

maxFreq = 3;  %
% Note: if maxFreq ≥ "f" of sine wave, then no distortion
% If maxFreq < "f" of input, then slew rate distortion
y = slewRateDistortion(x,Fs,maxFreq);

plot(t,x);hold on;
plot(t,y);hold off; axis([0 1 -1.1 1.1]);
legend('Input','Output');
```

Example 13.7: Script: `slewRateExample.m`

```
% SLEWRATEDISTORTION
% This function implements slew rate distortion. Frequencies
% greater than the "maxFreq" parameter are distorted.
```

```
% Frequencies less than "maxFreq" are not distorted. This type
% of distortion occurs in op-amps used for audio.
%
% Input Variables
%   in : input signal to be processed
%   Fs : sampling rate
%   maxFreq : the limiting/highest frequency before distortion

function [out] = slewRateDistortion(in,Fs,maxFreq)

Ts = 1/Fs;
peak = 1;
slewRate = maxFreq*2*pi*peak; % Convert freq to slew rate

slope = slewRate*Ts; % Convert slew rate to slope/sample

out = zeros(size(in));      % Total number of samples
prevOut = 0;         % Initialize feedback delay sample
for n = 1:length(in)

    % Determine the change between samples
    dlta = in(n,1) - prevOut;
    if dlta > slope  % Don't let dlta exceed max slope
        dlta = slope;
    elseif dlta < -slope
        dlta = -slope;
    end

    out(n,1) = prevOut + dlta;
    prevOut = out(n,1); % Save current "out" for next loop
end
```

Example 13.8: Function: `slewRateDistortion.m`

In Figure 13.27, slew rate distortion is shown for a sine wave input signal with a fundamental frequency greater than f_{max} for an op-amp. The result is an output signal resembling a triangle wave.

In Figure 13.28, slew rate distortion is shown for a square wave input signal with a fundamental frequency less than f_{max} for an op-amp. In this case, the result demonstrates the

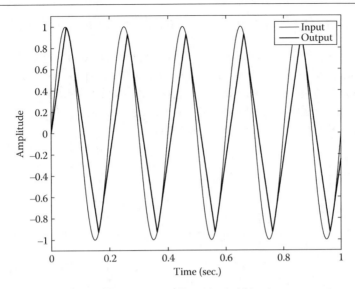

Figure 13.27: Slew rate distortion with a sine wave

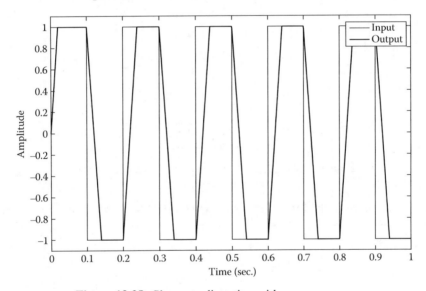

Figure 13.28: Slew rate distortion with a square wave

amplitude cannot abruptly transition between values for the square wave. Instead, the output resembles a trapezoidal wave.

Because the slew rate characteristic of an op-amp only causes distortion in frequencies greater than f_{max}, it can be described as a **frequency-dependent** distortion. This is different than the memoryless distortion effects presented in Chapter 10, which use element-wise processing and distort all frequencies in the same way.

13.9 Appendix III: Using an IIR Filter to Synthesize Pink Noise

In Appendix 12.9, the characteristics of pink noise are discussed, along with a process of synthesizing pink noise using an FIR filter. Here, the synthesis process is re-created using an IIR filter. In Example 13.9, a filter is implemented with a spectral slope of −3 dB per octave. The spectral magnitude response of the pink noise filter is shown in Figure 13.29, similar to the response of the FIR filter shown in Figure 12.34.

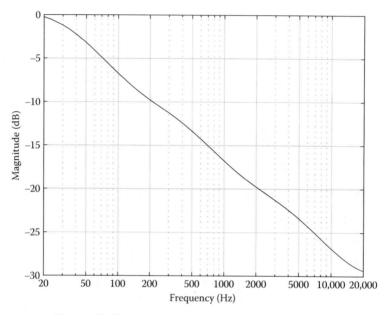

Figure 13.29: Magnitude response of pink noise filter

Examples: Create and run the following m-file in MATLAB.

```
% PINKNOISE2
% This script synthesizes an approximation of pink noise
% using an IIR filter.
%
% Pink noise can be created by filtering white noise. The
% amplitude response of the filter decreases by 10 dB/decade
% or -3 dB/octave.

clc; clear; close all;
Fs = 48000;      % Sampling rate
Nyq = Fs/2;      % Nyquist frequency for normalization
sec = 5;         % 5 seconds of noise
```

```
white = randn(sec*Fs,1);


% IIR coefficients
b = [0.049922035, -0.095993537, 0.050612699, -0.004408786];
a = [1 -2.494956002   2.017265875   -0.522189400];
[H,F] = freqz(b,a,4096,Fs);
semilogx(F,20*log10(abs(H))); axis([20 20000 -30 0]);


% Create pink noise by filtering white noise
pink = filter(b,a,white);
sound(pink,Fs);
```

Example 13.9: Script: `pinkNoise2.m`

13.10 Appendix IV: The LKFS/LUFS Standard Measurement of Loudness

A standard for measuring loudness of multi-channel audio signals has been established by the Radiocommunication Sector of the International Telecommunication Union (ITU-R BS.1770). The algorithm establishes a level of loudness, K-weighted, relative to full-scale amplitude (LKFS). This approach to measuring loudness has also been adopted by the European Broadcasting Union (EBU R128) as a loudness unit relative to full-scale amplitude (LUFS).

The algorithm for measuring audio loudness can be used for one to five channels. The process involves spectral filtering of each channel, followed by a calculation of mean-square energy, a gain weighting for each channel, and finally a summation of combined loudness. A block diagram of the two-channel case for stereo audio signals is shown in Figure 13.30.

An underlying premise of the algorithm is to perform an analysis that produces a result comparable to a listener's perception of loudness. Therefore, an approximation of a psychoacoustic model is used in the calculation.

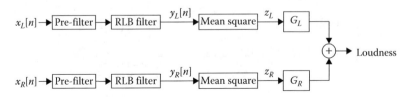

Figure 13.30: Block diagram of multichannel loudness algorithm

The first pre-filter is used to account for the acoustic effects of a listener's head and address the increased sensitivity of the auditory system to frequencies in the upper-mid range. Specifically, a high-shelf filter is used with a magnitude response shown in Figure 13.31.

A bi-quad filter is the recommended implementation of the pre-filter. The coefficients for a sampling rate of 48 kHz are listed in Table 13.2.

The pre-filter is followed by the revised low-frequency B (RLB) filter. It is based on a modified version of the B weighting curve (A, B, C weighting curves, DIN-IEC 651). It is a high-pass filter intended to reduce the sensitivity of the analysis to low frequencies. The magnitude response of the RLB filter is shown in Figure 13.32.

A bi-quad filter is the recommended implementation of the RLB filter. The coefficients for a sampling rate of 48 kHz are listed in Table 13.3.

The combined spectral response of the pre-filter and RLB filter is the K-weighting curve. Following the two stages of filtering, new signals for each channel are created, $y_L[n]$ and

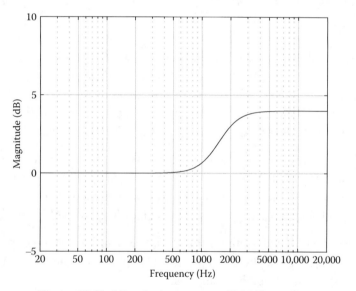

Figure 13.31: Magnitude response of LUFS pre-filter

Table 13.2: Bi-quad coefficients of LUFS pre-filter

b_0	1.53512485958697	a_0	1.0
b_1	−2.69169618940638	a_1	−1.69065929318241
b_2	1.19839281085285	a_2	0.73248077421585

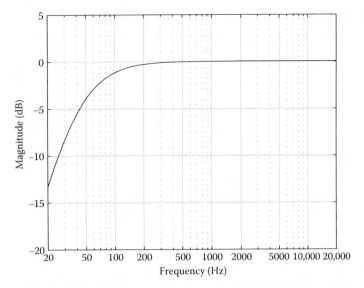

Figure 13.32: Magnitude response of LUFS RLB filter

Table 13.3: Bi-quad coefficients of LUFS RLB filter

b_0	1.0	a_0	1.0
b_1	−2.0	a_1	−1.99004745483398
b_2	1.0	a_2	0.99007225036621

$y_R[n]$, respectively. Then, the mean-square energy of each channel is measured:

$$z_i = \frac{1}{N} \sum_{n=1}^{N} (y_i[n])^2$$

where $N = \texttt{length(y)}$

and $i = $ L, R channels

Finally, the mean-square level for each channel is summed to calculate the value of LUFS:

$$\text{Loudness (dB)} = -0.691 + 10 \cdot log_{10}\left(G_L \cdot z_L + G_R \cdot z_R\right)$$

For stereo audio signals the scaling gains equal unity gain, $G_L = G_R = 1$. However, for other multi-channel audio signals, some scaling gains have other values.

Examples: Create and run the following m-files in MATLAB.

```
% LUFSEXAMPLE
% This script demonstrates the use of the lufs
% function for calculating loudness based
% on the LUFS/LKFS standard. Four examples
% are shown. First, a mono recording of
% an electric guitar is analyzed. Second,
% a stereo recording of drums is analyzed.
%
% Then, two more examples are demonstrated using
% test signals provided by the European Broadcast
% Union (EBU) for the sake of verifying proper
% measurement. A filtered pink noise signal
% is measured with a loudness of -23 LUFS. Finally,
% a sine wave signal is measured with a loudness
% of -40 LUFS.
%
% See also LUFS
clc;clear;

% Example 1 - Mono electric guitar
sig1 = audioread('RhythmGuitar.wav');
loudnessGuitar = lufs(sig1)

% Example 2 - Stereo drums
sig2 = audioread('stereoDrums.wav');
loudnessDrums = lufs(sig2)

% Example 3 - Filtered noise signal provide by EBU for verification
fnm='EBU-reference_listening_signal_pinknoise_500Hz_2kHz_R128.wav';
sig3 = audioread(fnm);
loudnessNoise = lufs(sig3)   % Should equal -23 LUFS

% Example 3 - Sine wave provided by EBU for verification
sig4 = audioread('1kHz Sine -40 LUFS-16bit.wav');
loudnessSine = lufs(sig4) % Should equal -40 LUFS
```

Example 13.10: Script: `lufsExample.m`

```
% LUFS
% This function calculates the loudness of a mono or stereo
% audio signal based on the LUFS/LKFS standard. The analysis
% involves multiple steps. First, the input signal is
% processed using the pre-filter followed by the RLB filter.
% Then a mean-square calculation is performed. Finally,
% all the channels are summed together and loudness is
% converted to units of decibels (dB).
%
% See also LUFSEXAMPLE

function [loudness] = lufs(x)
% Number of samples
N = length(x);
% Determine whether mono or stereo
numOfChannels = size(x,2);

% Initialize pre-filter
b0 = 1.53512485958697;
a1 = -1.69065929318241;
b1 = -2.69169618940638;
a2 = 0.73248077421585;
b2 = 1.19839281085285;
a0 = 1;

b = [b0, b1,b2];
a = [a0,a1,a2];

% Perform pre-filtering
w = zeros(size(x));
for channel = 1:numOfChannels % Loop in case it is stereo
    w(:,channel) = filter(b,a,x(:,channel));
end

% RLB filter
b0 = 1.0;
a1 = -1.99004745483398;
b1 = -2.0;
```

```
a2 = 0.99007225036621;
b2 = 1.0;
a0 = 1;
b = [b0, b1,b2];
a = [a0,a1,a2];


% Perform RLB filtering
y = zeros(size(x));
for channel = 1:numOfChannels
    y(:,channel) = filter(b,a,w(:,channel));
end


% Perform mean-square amplitude analysis
z = zeros(1,numOfChannels);
for channel = 1:numOfChannels
    % Add together the square of the samples,
    % then divide by the number of samples
    z(1,channel) = sum(y(:,channel).^2)/N;
end

% Determine loudness (dB) by summing all channels
loudness = -0.691 + 10 * log10(sum(z));
```

Example 13.11: Function: `lufs.m`

Bibliography

Bristow-Johnson, R. (2004, April 7). Cookbook formulae for audio EQ biquad filter coefficients. Retrieved from www.musicdsp.org/files/Audio-EQ-Cookbook.txt

European Broadcasting Union (2014). EBU R128 standard, *Loudness normalisation and permitted maximum level of audio signals* (pp. 3–5). Geneva: European Broadcasting Union. Retrieved from https://tech.ebu.ch/docs/r/r128.pdf

Radiocommunication Sector of ITU (2012). ITU-R BS.1770 standard, Annex 1 Specification of the objective multichannel loudness measurement algorithm (pp. 2–5). Geneva: International Telecommunication Union. Retrieved from www.itu.int/dms_pubrec/itu-r/rec/bs/R-REC-BS.1770-3-201208-S!!PDF-E.pdf

Chapter 14

Delay Buffers and Fractional Delay Interpolation
Linear and Circular Buffers, Linear and Cubic Interpolation

14.1 Introduction: Fractional Delay Using Buffers

This chapter is a digression from the topic of creating audio effects. Instead, it focuses on a general method of *how* time delay can be introduced digitally and implemented efficiently. This material is used for audio effects starting in Chapter 15.

In Chapters 11, 12, and 13, different approaches are presented for how delay effects can be created. First, array indexing at the current time sample can be used to reference the appropriate samples in the past based on the delay time. Second, the MATLAB function conv can be used with the impulse response of an FIR system. Third, the feedforward and feedback gain coefficients can be used with the MATLAB function filter to implement an IIR system.

These three methods are sufficient for time-invariant systems used in audio effects. But for the practical implementation of time-variant modulated delay effects, these methods cannot be used directly. Furthermore, modulated audio effects do not always use a delay time based on a whole number of samples. Instead, interpolated fractional delays are used. All these things provide motivation for an approach of creating fractional delay effects using buffers.

14.2 Delay Buffers

An intuitive method for implementing time-variant (and also time-invariant) delay involves using **delay buffers**. A delay buffer is an allocated array used as memory to store previous values of a signal.

273

x_n	x_{n-1}	x_{n-2}	x_{n-3}	x_{n-4}	$\cdots$	x_{n-d}

Figure 14.1: Linear delay buffer

Initial :	0	0	0	0	0	0	0
$n = 1$:	x_1	0	0	0	0	0	0
$n = 2$:	x_2	x_1	0	0	0	0	0
$n = 3$:	x_3	x_2	x_1	0	0	0	0
$n = 4$:	x_4	x_3	x_2	x_1	0	0	0
$\vdots$							
$n = 7$:	x_7	x_6	x_5	x_4	x_3	x_2	x_1
$n = 8$:	x_8	x_7	x_6	x_5	x_4	x_3	x_2

Figure 14.2: Iterations of a buffer to create delay

14.2.1 Linear Buffers

One approach to using a delay buffer is as a **linear buffer**. In this case, the current signal value is stored in the first element of the array, while iterating through the individual samples in a signal. The signal value with one sample of delay is stored in the second element of the array, and the signal value with two samples of delay is stored in the third element of the array, etc. Therefore, the number of samples of delay is relative to the first element of the array. For sample number n, the organization of the samples in signal x within a delay buffer is shown in Figure 14.1.

During each iteration, the values in each element of the buffer are shifted to make space for the current signal value to be stored in the first element. Consider the example in Figure 14.2 for the signal $x = \{x_1, x_2, x_3, x_4, x_5, x_6, x_7, x_8\}$. Every value in the delay buffer is initialized to zero.

Elements from the array can be indexed to read the delayed values at the appropriate time. During an iteration one of the elements of the delay buffer is stored in the current sample of the output signal, and the current sample of the input signal is stored in the first element of the delay buffer. Then all the elements of the delay buffer are shifted and the process repeats. An example of this process is shown in Figure 14.3.

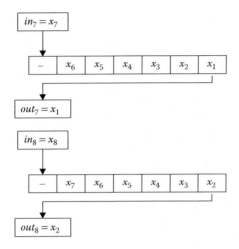

Figure 14.3: Indexing a delay buffer

An implementation of a linear buffer is demonstrated in Example 14.1.

Examples: Create and run the following m-file in MATLAB.

```
% SIMPLELINEARBUFFER
% This script demonstrates the basics of creating a linear buffer.
% An input signal is processed by a loop to index the
% "current" sample and store it in the delay buffer. Each time
% through the loop the delay buffer is shifted to make room
% for a new sample. The output of the process is determined
% by indexing an element at the end of the delay buffer.
%
% Execution of the script is set to "pause" during each iteration
% to allow a user to view the contents of the delay buffer
% during each step. To advance from one step to the next,
% press <ENTER>
%
% See also DELAYBUFFEREXAMPLE
clear;clc;
% Horizontal for displaying in Command Window
in = [1, -1, 2 , -2, zeros(1,6)];

% Buffer should be initialized without any value
```

```
% Length of buffer = 5, output is indexed from end of buffer
% Therefore, a delay of 5 samples is created
buffer = zeros(1,5);


N = length(in);
out = zeros(1,N);
for n = 1:N

    % Read the output at the current time sample
    % from the end of the delay buffer
    out(1,n) = buffer(1,end);
    disp(['For sample ', num2str(n),' the output is: ', ...
        num2str(out(1,n))]);

    % Shift each value in the buffer by one element
    % to make room for the current sample to be stored
    % in the first element
    buffer = [in(1,n)  buffer(1,1:end-1)];
    disp(['For sample ', num2str(n),' the buffer is: ', ...
        num2str(buffer)]);

    % Press <ENTER> in the Command Window to continue
    pause;
    disp([' ']); % Blank line
end


% Compare the input & output signals
disp(['The orig. input signal was: ', num2str(in)]);
disp(['The final output signal is: ', num2str(out)]);
```

Example 14.1: Script: `simpleLinearBuffer.m`

The process in Example 14.1 can be modified to implement several different types of systems including a series delay with arbitrary delay length, feedforward parallel delay, and feedback delay. Each type of system is demonstrated in Example 14.2.

Examples: Create and run the following m-file in MATLAB.

```
% DELAYBUFFEREXAMPLES
% This script demonstrates several examples of creating different
% types (FIR, IIR) of systems by using a delay buffer
```

```
%
% Execution of the script is set to "pause" during each iteration
% to allow a user to view the contents of the delay buffer
% during each step. To advance from one step to the next.
% press <ENTER>
%
% See also SIMPLELINEARBUFFER
clear;clc;
in = [1, -1, 2 , -2, zeros(1,6)];


buffer = zeros(1,20); % Longer buffer than delay length

% Number of samples of delay
delay = 5; % Does not need to be the same length as buffer


N = length(in);
out = zeros(1,N);
% Series delay
for n = 1:N

    out(1,n) = buffer(1,delay);
    buffer = [in(1,n) , buffer(1,1:end-1)];

end

% Compare the input & output signals
disp('Series Delay: 5 Samples');
disp('out(n) = in(n-5)');
disp(['The orig. input signal was: ', num2str(in)]);
disp(['The final output signal is: ', num2str(out)]);
pause; clc; % Press <ENTER> to continue


% Feedforward (FIR) system
buffer = zeros(1,20);
delay = 3; % Number of samples of delay

% Parallel delay line
for n = 1:N
```

```
    out(1,n) = in(1,n) + buffer(1,delay);
    buffer = [in(1,n) , buffer(1,1:end-1)];

end

disp('Feed-forward Delay: 3 samples');
disp('out(n) = in(n) + in(n-3)');
disp(['The orig. input signal was: ', num2str(in)]);
disp(['The final output signal is: ', num2str(out)]);
pause; clc; % Press <ENTER> to continue


% Feedback (IIR) system
buffer = zeros(1,20);
for n = 1:N

    out(1,n) = in(1,n) + buffer(1,delay);
    buffer = [out(1,n) , buffer(1,1:end-1)];

end

disp('Feed-back Delay: 3 samples');
disp('out(n) = in(n) + out(n-3)');
disp(['The orig. input signal was: ', num2str(in)]);
disp(['The final output signal is: ', num2str(out)]);
```

Example 14.2: Script: delayBufferExample.m

Using Delay Buffers in MATLAB Functions

It is possible to use delay buffers within functions for an audio effect. Here, a single sample of the input signal along with the delay buffer are passed as input variables to the function. Then, both the output sample and the iterated, shifted delay buffer are returned from the function for future use.

Examples: Create and run the following m-files in MATLAB.

```
% FEEDBACKDELAYEXAMPLE
% This script calls the feedback delay function and passes in
% the delay buffer
%
```

```
% See also FEEDBACKDELAY


clear;clc;
in = [1 ; -1 ; 2 ; -2 ; zeros(6,1)]; % Input signal


% Longer buffer than delay length to demonstrate
% delay doesn't just have to be the "end" of buffer
buffer = zeros(20,1);


% Number of samples of delay
delay = 5;


% Feedback gain coefficient
fbGain = 0.5;


% Initialize output vector
N = length(in);
out = zeros(N,1);


% Series delay
for n = 1:N
    % Pass "buffer" into feedbackDelay function
    [out(n,1),buffer] = feedbackDelay(in(n,1),buffer,delay,fbGain);
    % Return updated "buffer" for next loop iteration
end
% Print and compare input and output signals
disp('Feed-back Delay: 5 samples');
disp('The orig. input signal was: ')
in
disp('The final output signal is: ');
out
```

Example 14.3: Script: `feedbackDelayExample.m`

```
% FEEDBACKDELAY
% This function performs feedback delay by processing an
% individual input sample and updating a delay buffer
% used in a loop to index each sample in a signal
%
% Additional input variables
%        delay : samples of delay
```

```
%        fbGain : feedback gain (linear scale)

function [out,buffer] = feedbackDelay(in,buffer,delay,fbGain)

out = in + fbGain*buffer(delay,1);

% Store the current output in appropriate index
buffer = [out ; buffer(1:end-1,1)];
```

Example 14.4: Function: `feedbackDelay.m`

The process of shifting every element one position during each iteration can be relatively computationally complex, especially if the delay buffer has many values. In the shifting process, it is necessary to read the value from one element and write it to a different element. This is done as many times as the number of samples stored in the buffer. This process is shown for a single iteration in Figure 14.4.

14.2.2 Circular Buffers

One approach to avoid the inefficient task of shifting samples in a linear buffer is to use a **circular buffer**. In a circular buffer, the value of each element is stored one time in its location and does not shift during an iteration. Instead, the relative location of the start of the buffer is changed during an iteration. Therefore, it is only necessary to change one value (the

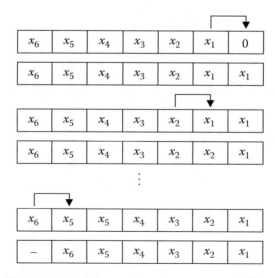

Figure 14.4: Shifting the elements in a linear buffer

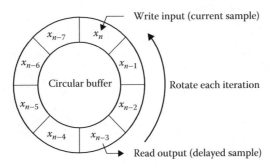

Figure 14.5: Indexing a circular delay buffer

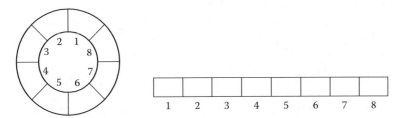

Figure 14.6: Relationship between a circular buffer and an array

buffer index) during an iteration instead of every element of the buffer. The index of the delayed sample is determined relative to the current buffer index.

The visualization of a circular buffer as a round array is a convenient way to explain the cyclical repetition of indexing, shown in Figure 14.5. In practice, a computer does not actually arrange data in circles. Instead, a programmer stores data in a conventional array and indexes it cyclicly. Each element of an array is indexed from the first to the last. After the last element, the index returns to the start of the array to repeat again. An example is shown in Figure 14.6 of the element indexes for a circular buffer and an array.

Consider the example in Figure 14.7 for signal $x = \{x_1, x_2, x_3, x_4, x_5, x_6, x_7, x_8, x_9\}$. During an iteration, a single sample of the signal is stored at the current index, and a separate index can be used to read a delayed output signal. During the next iteration, the current index changes and the next sample is stored. This process repeats until the indexing completes a cycle and starts over again from the original index.

Modulo Operator

For cyclically indexing an array, it is necessary to use a method that does not index outside the dimensions of the array. The built-in function mod can be used to implement this method.

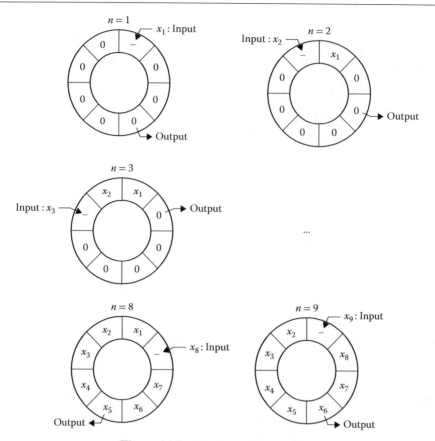

Figure 14.7: Indexing a delay buffer

The function mod calculates the remainder after division for two numbers. This is called the modulo operator. For numbers a and m, where a is the dividend and m is the divisor (i.e., $\frac{a}{m}$), mod(a,m) calculates a - m * floor(a/m).

When used as part of a loop that counts whole numbers, the result of the modulo operator repeats cycles bounded by the value of m.

Examples: Create and run the following m-file in MATLAB.

```
% MODULOOPERATOR
% This script demonstrates the result of the modulo
% operator - mod(a,m);
%
% See also MOD

clear;clc;
```

```
m = 4;
for a = 1:15

    disp(['When "a" is : ', num2str(a)]);
    disp(['mod(a,4) is : ', num2str(mod(a,m))]); disp(' ');

end
```

Example 14.5: Script: `moduloOperator.m`

Therefore, the modulo operator can be used to determine the current index for a circular buffer based on the sample number. Then, a delay can be created using the circular buffer by indexing a different element of the array relative to the current index.

Examples: Create and run the following m-files in MATLAB.

```
% CIRCULARBUFFEREXAMPLE
% This script tests a circular buffer function and
% demonstrates how it works
%
% Execution of the script is set to "pause" during each iteration
% to allow a user to view the contents of the delay buffer
% during each step. To advance from one step to the next,
% press <ENTER>
%
% See also CIRCULARBUFFER

clear;clc;
in = [1 ; -1 ; 2 ; -2 ; 3 ; zeros(5,1)];
buffer = zeros(6,1);

% Number of samples of delay
delay = 4;

N = length(in);
out = zeros(N,1);
% Series delay
for n = 1:N

    [out(n,1),buffer] = circularBuffer(in(n,1),buffer,delay,n);
    % Display current status values
```

```
        disp(['The current sample number is: ' , ...
            num2str(n)]);
        disp(['The current buffer index is: ' , ...
            num2str(mod(n-1,6) + 1)]);
        disp(['The current delay index is: ' , ...
            num2str(mod(n-delay-1,6) + 1)]);
        disp(['The input is: ' , num2str(in(n,1))]);
        disp(['The delay buffer is: [' , num2str(buffer.') , ']']);
        disp(['The output is: ' , num2str(out(n,1))]); disp(' ');
        pause;

end
```

Example 14.6: Script: `circularBufferExample.m`

```
% CIRCULARBUFFER
% This function performs series delay and uses
% a circular buffer. Rather than shifting
% all the values in the array buffer during each
% iteration, the index changes each time through
% based on the current sample, "n"
%
% Additional input variables
%    delay : samples of delay
%    n : current sample number used for circular buffer

function [out,buffer] = circularBuffer(in,buffer,delay,n)

% Determine indexes for circular buffer
len = length(buffer);
indexC = mod(n-1,len) + 1; % Current index in circular buffer
indexD = mod(n-delay-1,len) + 1; % Delay index in circular buffer

out = buffer(indexD,1);
% Store the current output in appropriate index
buffer(indexC,1) = in;
```

Example 14.7: Function: `circularBuffer.m`

Delay buffers provide an generalized approach for implementing all types of digital delay. This includes delay used in time-variant modulation audio effects, which require fractional delay.

14.3 Fractional Delay

Some audio effects have a modulated delay time. These effects could be implemented using delay by a whole number of samples. Over time, the number of samples of delay changes. For instance, a delay of six samples could be used for part of a signal followed immediately by a delay of seven samples for the next part of the signal. However, by having a relatively large discontinuity to change from six samples to seven samples of delay, an undesirable audible distortion is introduced.

For modulated delay effects, it is desirable for the delay time to change smoothly. Ideally, the discontinuity distortion would be inaudible. To gradually change delay time, it is necessary to use **fractional delay**, instead of a whole number of samples. The concept is to make it possible to transition smoothly from a delay of n to $n+1$ samples. As an example, the delay time could change based on the following values: $6.0, 6.1, 6.2, 6.3, ..., 6.8, 6.9, 7.0$.

A digital signal's amplitude is undefined (i.e., not represented) between samples. Fractional delay approximates a value of the underlying, continuous signal between samples. It is based on estimating the value between samples by using **interpolation**. There are many methods of interpolation, each providing one approximation of the underlying, continuous signal (Niemitalo, 2001). Linear and cubic interpolation are presented here as a conceptual starting point for exploring the various options.

14.3.1 Linear Interpolation

For linear interpolation, the value of a signal between samples is estimated by a linear combination of the two closest samples. For a signal, $x = \{x_1, x_2, x_3, x_4, ..., x_n\}$, a value between samples m and $m+1$ is estimated using the following expression:

$$(1 - f_{rac}) \cdot x_m + (f_{rac}) \cdot x_{m+1}$$

where f_{rac} is a fraction of a time sample, $0 < f_{rac} < 1$. This method of interpolation estimates values along a straight line connecting between adjacent samples. Although the underlying, continuous signal is not actually a straight line between samples, linear interpolation can be a sufficient approximation in many situations. An example of linearly interpolated values for a signal is shown in Figure 14.8.

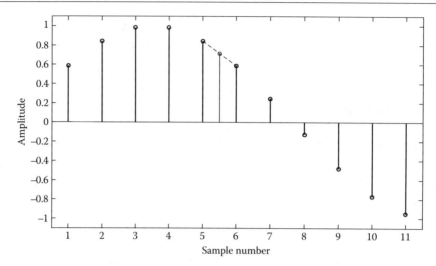

Figure 14.8: Linear interpolation example

Linear interpolation can be implemented to perform a fractional delay by indexing sequential samples in a delay buffer. Rather than a linear combination of the current sample, the delay is accomplished by combining together samples stored in the buffer at the appropriate index number. A demonstration of the fractional delay is shown in Example 14.8.

Examples: Create and run the following m-file in MATLAB.

```
% LINEARINTERPOLATIONDELAY
% This script demonstrates how to introduce a
% fractional (non-integer) delay. Linear interpolation
% is used to estimate an amplitude value in between
% adjacent samples.

clear;clc;
in = [1 , zeros(1,9)]; % Horizontal for displaying in command window

fracDelay = 3.2;           % Fractional delay length in samples
intDelay = floor(fracDelay); % Round down to get the previous (3)
frac = fracDelay - intDelay; % Find the fractional amount (0.2)

buffer = zeros(1,5); % length(buffer) ≥ ceil(fracDelay)

N = length(in);
```

```
out = zeros(1,N);
% Series Fractional Delay
for n = 1:N

    out(1,n) = (1-frac) * buffer(1,intDelay) + ...
        (frac) * buffer(1,intDelay+1);

    buffer = [in(1,n)  buffer(1,1:end-1)];
end

% Compare the input & output signals
disp(['The orig. input signal was: ', num2str(in)]);
disp(['The final output signal is: ', num2str(out)]);
```

Example 14.8: Script: linearInterpolationDelay.m

Feedforward LPF

Another way to understand linear fractional delay is to consider the basic feedforward LPF presented in Section 12.2.1. A block diagram of the system with arbitrary gain values, b_0 and b_1, is shown in Figure 14.9.

Consider two trivial cases. First, if $b_0 = 1$ and $b_1 = 0$, the system passes the input signal through with 0 samples of delay. Second, if $b_0 = 0$ and $b_1 = 1$, the system passes the input signal through with 1 sample of delay. Equivalently, 1 sample of delay has been shown to introduce a phase shift, ϕ, in radians for a frequency, f, in Hz.

Next, consider the case when $b_0 = b_1 = 0.5$, discussed in Section 12.2.1. The system output for a given frequency can be written in the form: $A \cdot sin(x + \frac{\phi}{2})$. Since the phase shift, ϕ, results from 1 sample of delay, then the phase shift, $\frac{\phi}{2}$, is equivalent to 0.5 samples of delay.

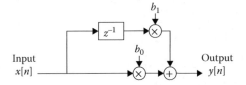

Figure 14.9: Block diagram of LPF to create fractional delay

Furthermore, by modifying the gains in a complementary way between 0 and 1, the LPF can be used to perform linear interpolation to estimate any delay time between 0 and 1 samples. However, this also illustrates a drawback of linear interpolation. It acts as a LPF to decrease the amplitude of high frequencies. Other approaches to interpolation can be used to ameliorate this problem.

Group Delay

As shown in the previous example, a system may have several paths, with various delay lengths, all combined together. The result is an output signal which may have a phase shift relative to the input signal. The phase shift of the system can be converted to a delay in samples of the system, which is called **group delay**.

Group delay is calculated by using the slope of the phase shift between two frequencies, f_1 and f_2. For a phase shift in units of degrees, the group delay, $\tau_d = \frac{\phi_2-\phi_1}{f_2-f_1} \cdot \frac{1}{-180°}$. For a phase shift in units of radians, the group delay, $\tau_d = \frac{\phi_2-\phi_1}{f_2-f_1} \cdot \frac{1}{-\pi}$.

For the basic LPF in Figure 14.9 with $b_0 = b_1 = 0.5$, the slope of the phase shift is constant between normalized frequencies 0 an 1. Therefore, the group delay can be calculated:

$$\tau_d = \frac{-90° - 0°}{1-0} \cdot \frac{1}{-180°} = 0.5 \text{ samples}$$

In general, systems with a linear phase response introduce a constant group delay for all frequencies (e.g. all frequencies are delayed 0.5 samples).

However, systems which do not have a linear phase response introduce a frequency-dependent group delay. Therefore, the slope should not be measured across the entire spectrum of normalized frequencies. Instead, the slope should be measured between adjacent discrete frequencies to determine the group delay for a given frequency.

MATLAB Filter Visualization Tool

The Signal Processing Toolbox has a tool for visualizing filter characteristics including group delay. The syntax for using the tool is: `fvtool(b,a)`, for a system defined by feed-forward and feed-back coeffcients, b and a, respectively.

The magnitude response, phase response, and group delay can be viewed by clicking the button on the *Filter Visualization* figure window for each analysis. In Figure 14.10, the LPF created by the Butterworth filter design function, `[b,a] = butter(2,0.25)`, is visualized using `fvtool(b,a)`.

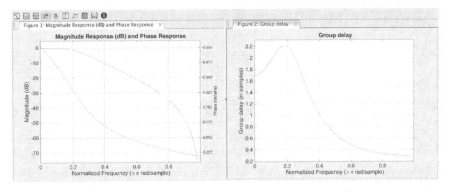

Figure 14.10: Visualizing group delay using `fvtool`

14.3.2 Cubic Interpolation

Another type of interpolation is called **cubic interpolation**. It is based on a polynomial equation of the third order. Cubic interpolation determines an estimated value of a signal between two samples by incorporating the amplitude of the four closest samples. In theory, by incorporating additional samples in the estimate, then a more accurate estimate can be made. However, an increase in computational complexity is necessary, as each estimate requires additional processing steps.

To perform cubic interpolation for a signal, $x = \{x_1, x_2, x_3, x_4, ..., x_n\}$, a value between samples m and $m+1$ is estimated using the following steps:

$$\text{Let: } a_0 = x_{m+2} - x_{m+1} - x_{m-1} + x_m$$

$$a_1 = x_{m-1} - x_m - a_0$$

$$a_2 = x_{m+1} - x_{m-1}$$

$$a_3 = x_m$$

$$y = a_0 \cdot (f_{rac}^3) + a_1 \cdot (f_{rac}^2) + a_2 \cdot f_{rac} + a_3$$

where f_{rac} is a fraction of a time sample, $0 < f_{rac} < 1$ after x_m.

Examples: Create and run the following m-file in MATLAB.

```
% CUBICINTERPOLATIONDELAY
% This script demonstrates how to introduce a
% fractional (non-integer) delay. Cubic interpolation
% is used to estimate an amplitude value in-between
% adjacent samples.
```

```matlab
clear;clc;
in = [1 , zeros(1,9)]; % Horizontal for displaying in Command Window

fracDelay = 3.2;              % Fractional delay length in samples
intDelay = floor(fracDelay); % Round down to get the previous (3)
frac = fracDelay - intDelay; % Find the fractional amount (0.2)

buffer = zeros(1,5); % length(buffer) ≥ ceil(fracDelay)+1

N = length(in);
out = zeros(1,N);
% Series fractional delay
for n = 1:N
    % Calculate intermediate variable for cubic interpolation
    a0 = buffer(1,intDelay+2) - buffer(1,intDelay+1) - ...
        buffer(1,intDelay-1) + buffer(1,intDelay);

    a1 = buffer(1,intDelay-1) - buffer(1,intDelay) - a0;

    a2 = buffer(1,intDelay+1) - buffer(1,intDelay-1);

    a3 = buffer(1,intDelay);

    %
    out(1,n) = a0*(frac^3) + a1*(frac^2) + a2*frac + a3;

    buffer = [in(1,n)  buffer(1,1:end-1)];     % Shift buffer
end

% Compare the input & output signals
disp(['The orig. input signal was: ', num2str(in)]);
disp(['The final output signal is: ', num2str(out)]);

plot(out);
% Observe in this plot that the impulse at sample n=1 is delayed
% by 3.2 samples. Therefore, the output signal should have
% an impulse at time 4.2 samples. With cubic interpoloation
% this impulse contributes to the amplitude of the output
% signal at samples 3,4,5,6. The result of cubic interpolation
```

```
% is a closer approximation to the underlying (smooth)
% continuous signal than linear interpolation.
```

Example 14.9: Script: `cubicInterpolationDelay.m`

Many other methods exist for performing interpolation, but these are not discussed here. Alternative methods of interpolation can be substituted in an algorithm to best fulfill design specifications. To generalize the approach from linear interpolation, and extended with cubic interpolation, an estimate of a signal can be made between time samples by combining together the values of several of the adjacent samples in different ways.

Bibliography

Niemitalo, O. (2001, October). Polynomial interpolators for high-quality resampling of oversampled audio. Retrieved from http://yehar.com/blog/wp-content/uploads/2009/08/deip.pdf

Chapter 15

Modulated Delay Effects
Vibrato, Chorus, Flanger, Pitch Shifter, and Phaser

15.1 Introduction: Modulated Delay Effects

This chapter introduces several different audio effects which use a time-varying, or modulated, delay. These effects expand on the delay effects described in Chapters 11, 12, and 13, while making use of the delay buffers and fractional delay processing discussed in Chapter 14. Furthermore, the individual effects presented in this chapter are used in combination as the processing blocks for algorithmic reverb in Chapter 16.

15.1.1 Time-Invariant and Time-Variant Systems

Many audio effects perform the same processing to a signal at all times. These systems do not change over time and are described as time-invariant systems. In this case, the routing, delay, amplitude coefficients, and any other processing is applied to the first sample and the last sample in the exact same way.

There are other audio effects that are created using time-variant systems. In this case, one or more parameters of the system change over time. Using the tremolo effect as an example, one sample in a signal is processed by one amplitude change, while the next sample in the signal is processed by a different amplitude change. For modulated delay effects, one sample in a signal is delayed by one delay amount and a different sample in the signal is delayed by a different amount.

15.1.2 Types of Modulated Delay Effects

There are several common audio effects that can be created using a modulated delay. Examples of these effects include vibrato, chorus, flanger, pitch shifter, and phaser. Each type is created by changing whether a time delay is used as a series or parallel effect with either a relatively short or long delay time.

Before discussing how each type of modulated delay effect is created, one method for analyzing the effects, called a spectrogram, is presented.

15.2 Visualizing Time-Varying Information

For modulated delay effects, it is necessary to analyze how the system, or a processed signal by the effect, varies over time. In several of the effects, there are spectral changes happening over time, either in a signal's pitch or in the important frequencies of a filter.

15.2.1 Spectrogram

A **spectrogram** is a three-dimensional visualization of a signal's amplitude over frequency and time. Many audio signals are comprised of multiple frequencies happening at the same time, and are also signals that change to different frequencies at different times. Therefore, the spectrogram is a common way for audio engineers to analyze a complex signal. The Signal Processing Toolbox has a function called `spectrogram` that can be used to perform this analysis.

Signal Segmentation

To create a spectrogram, the signal's waveform is segmented into short time frames. Then the spectral content of each short frame can be analyzed and compared. Typically, the number of samples included in a single time frame is selected to be a power of 2. As an example, the spectral content of the first 2048 samples of a signal is plotted alongside the spectral content of the next 2048 samples, followed by the next time frame, etc.

As a rule of thumb, by using a longer time frame, increased frequency resolution can be achieved at the expense of decreased temporal resolution. By using a shorter time frame, increased temporal resolution can be achieved at the expense of decreased frequency resolution.

In some cases, the segmentation is performed to have an overlap of some number of samples shared between sequential frames. Additionally, a windowing function (Hann, Blackman, Hamming, etc.) can be used to smooth over the transition between overlapping frames. These parameters can be used with the MATLAB `spectrogram` function, but their details are not discussed here. In Example 15.1, an audio signal is analyzed using the waveform and spectrogram plots. The result is shown in Figure 15.1.

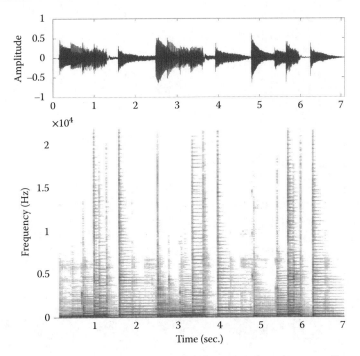

Figure 15.1: Spectrogram example of an acoustic guitar recording

Examples: Create and run the following m-file in MATLAB.

```
% AUDIOSPECGRAM
% This script displays a spectrogram of an audio signal
% using the built-in MATLAB function
%
% See also SPECTROGRAM


[x,Fs] = audioread('AcGtr.wav');


% Waveform
t = [0:length(x)-1].' * (1/Fs);
subplot(3,1,1);
plot(t,x); axis([0 length(x)*(1/Fs) -1 1]);


% Spectrogram
nfft = 2048; % Length of each time frame


window = hann(nfft); % Calculated windowing function
```

```
overlap = 128; % Number of samples for frame overlap

% Use the built-in spectrogram function
[y,f,t,p] = spectrogram(x,window,overlap,nfft,Fs);

% Lower subplot
subplot(3,1,2:3);
surf(t,f,10*log10(p),'EdgeColor','none');
% Rotate the spectrogram to look like a typical "audio" visualization
axis xy; axis tight; view(0,90);
xlabel('Time (sec.)');ylabel('Frequency (Hz)');
```

Example 15.1: Script: `audioSpecGram.m`

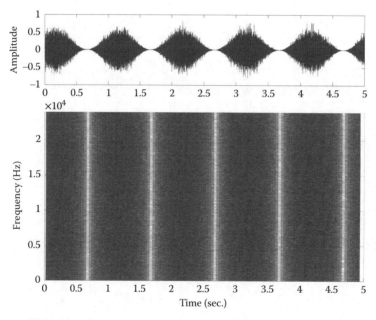

Figure 15.2: Waveform and spectrogram of amplitude modulated white noise

As an additional example for visualizing signals using different plots, the waveform and spectrogram for amplitude-modulated white noise is shown in Figure 15.2.

15.3 Low-Frequency Oscillator

The delay time for a modulated delay effect is typically controlled by a low-frequency oscillator (LFO). This type of signal is discussed in Section 8.5 for its use in amplitude modulation effects. In that case, the amplitude of the LFO was used to change the amplitude of the input carrier signal. For modulated delay effects, the amplitude of the LFO is used to change the delay. If the amplitude of the LFO increases, then the delay time increases. If the amplitude of the LFO decreases, then the delay time decreases.

As an example, consider a sine-wave LFO based on the following function:

$$3 \cdot sin(2 \cdot \pi \cdot f \cdot t) + 5$$

Here, the amplitude of the sine wave is offset, such that it is oscillating with an amplitude range centered around a value of 5. Therefore, this LFO could be used to create a delay time centered around 5 samples (or could be converted to other time units, 5 milliseconds or 5 seconds). The amplitude of the sine wave increases to a maximum of 8 and decreases to a minimum of 2. As the amplitude changes, this can be used to change the delay time between 8 samples and 2 samples. Typically, the frequency, f, of the LFO for musical effects can range from 0.1–10 Hz.

15.4 Series Delay

One method to incorporate a modulated delay as an effect uses series processing. Here, the input signal is routed though the delay and only the processed signal is used for the output. A block diagram for the series modulated delay effect is shown in Figure 15.3.

15.4.1 Vibrato Effect

The series modulated delay is called **vibrato** when used as an audio effect. Although this effect is based on time delay, this effect is not typically used to create an audible echo. Instead, vibrato introduces slight changes to the pitch of a signal when the delay is increasing or decreasing. When the delay is increasing (i.e., getting longer), it is as if the signal is slowing down. Therefore, the period of the signal is elongated and the frequency decreases. When the delay is decreasing (i.e., getting shorter), it is as if the signal is speeding up. Therefore, the period of the signal is shortened and the frequency increases (Dutilleux and Zölzer, 2002).

Doppler Effect

A related concept found in acoustic waves is the **Doppler effect**. When a sound source is moving towards an observer, the frequency of the signal increases because the wavelength

Figure 15.3: Block diagram of modulated delay in series

gets compressed. When a sound source is moving away from an observer, the frequency of the signal decreases because the wavelength is expanded.

Vibrato Parameters

In practice, the vibrato effect typically goes through a pattern of increasing and decreasing frequency. The changes to the effect are controlled by an LFO. The amplitude and frequency of the LFO are parameters for modulating the delay time. These parameters are commonly labeled as *depth* (amplitude) and *rate* (frequency). Conceptually, the depth parameter is used to adjust how wide of a range the pitch is changed and the rate is used to adjust how fast the pitch is changed.

Examples: Create and run the following m-files in MATLAB.

```
% VIBRATOEXAMPLE
% This script creates a vibrato effect, applied to an acoustic
% guitar recording. Parameters for the effect include "rate"
% and "depth," which can be used to control the intensity
% of the vibrato. At the end of the script, the sound of
% the result is played.
%
% See also VIBRATOEFFECT
clear;clc;

[in,Fs] = audioread('AcGtr.wav'); % Input signal
Ts = 1/Fs;
N = length(in);

% Initialize the delay buffer
maxDelay = 1000; % Samples
buffer = zeros(maxDelay,1);

% LFO parameters
t = [0:N-1]*Ts; t = t(:);
rate = 4; % Frequency of LFO in Hz
depth = 75; % Range of samples of delay

% Initialize output signal
out = zeros(N,1);
```

```
for n = 1:N

    [out(n,1),buffer] = vibratoEffect(in(n,1),buffer,Fs,n,...
      depth,rate);

end

sound(out,Fs);
```

Example 15.2: Script: `vibratoExample.m`

```
% VIBRATOEFFECT
% This function implements a vibrato effect based
% on depth and rate LFO parameters.
%
% Input Variables
%   in : single sample of the input signal
%   buffer : used to store delayed samples of the signal
%   n : current sample number used for the LFO
%   depth : range of modulation (samples)
%   rate : speed of modulation (frequency, Hz)
%
% See also CHORUSEFFECT

function [out,buffer] = vibratoEffect(in,buffer,Fs,n,...
    depth,rate)

% Calculate lfo for current sample
t = (n-1)/Fs;
lfo = (depth/2) * sin(2*pi*rate*t) + depth;

% Determine indexes for circular buffer
len = length(buffer);
indexC = mod(n-1,len) + 1; % Current index in circular buffer

fracDelay = mod(n-lfo-1,len) + 1; % Delay index in circular buffer
intDelay = floor(fracDelay);       % Fractional delay indices
frac = fracDelay - intDelay;

nextSamp = mod(intDelay,len) + 1; % Next index in circular buffer
```

```
out = (1-frac) * buffer(intDelay,1) + ...
    (frac) * buffer(nextSamp,1);

% Store the current output in appropriate index
buffer(indexC,1) = in;
```

Example 15.3: Function: `vibratoEffect.m`

15.5 *Parallel Delay*

An alternative method to incorporate a modulated delay in an effect is to use parallel processing. In this case, an unprocessed dry version of the input signal is blended with the processed wet version of the signal. A blending gain control, g, is used to adjust the relative level of the dry and wet signals. For some audio effects, the blend is used to control the perceived level of the delayed echoes. In other effects, the blend can be thought of as the control for the amount of constructive and destructive interference created by combining the parallel paths. A block diagram for the parallel modulated delay effect is shown in Figure 15.4.

15.5.1 *Chorus Effect*

One audio processor created using a parallel modulated delay is the **chorus** effect. Conceptually, the chorus effect is meant to process an individual voice and make it sound like a choir is performing the same part. This effect is used on many different types of instruments and signals, in addition to singing. In general, chorus is meant to simulate the sound of multiple musicians performing the same part. It is as if a part has been double- or triple-tracked (i.e., recorded two or three times) and layered to add a perceived "thickening" to the sound.

When multiple people sing the same part, there are inevitably small differences between the signals they produce. Each singer may start and stop notes at slightly different times. The

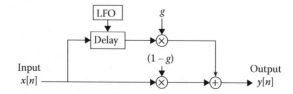

Figure 15.4: Block diagram of feedforward modulated delay

pitch of each note sung by the singers might be a fraction of a semitone different. Both of these differences create a "thickening" effect and give a choir its sound.

The chorus effect is closely related to the vibrato effect (Section 15.4.1). Chorus is created by blending the unprocessed dry input signal with the vibrato effect in parallel. The delay time of the effect typically has a range between 10–50 ms. This range is close to a listener's echo threshold (Section 11.4), such that the effect does not create an audible echo due to a longer time delay.

Chorus Parameters

The parameters for the chorus effect are similar to the vibrato effect. *Rate* and *depth* adjust the amplitude and frequency, respectively, of the LFO used to modulate delay time. Additionally, a *pre-delay* parameter can be used as an amplitude offset such that the delay time is within a range of 10–50 ms.

Examples: Create and run the following m-files in MATLAB.

```matlab
% CHORUSEXAMPLE
% This script creates a chorus effect, applied to an acoustic
% guitar recording.
%
% See also CHORUSEFFECT
clear;clc;

[in,Fs] = audioread('AcGtr.wav');

maxDelay = ceil(.05*Fs);  % Maximum delay of 50 ms
buffer = zeros(maxDelay,1);

rate = 0.6; % Hz (frequency of LFO)
depth = 5; % Milliseconds (amplitude of LFO)
predelay = 30; % Milliseconds (offset of LFO)

wet = 50; % Percent wet (dry = 100 - wet)

% Initialize output signal
N = length(in);
out = zeros(N,1);
```

```
for n = 1:N


    % Use chorusEffect.m function
    [out(n,1),buffer] = chorusEffect(in(n,1),buffer,Fs,n,...
        depth,rate,predelay,wet);

end


sound(out,Fs);
```

Example 15.4: Script: `chorusExample.m`

```
% CHORUSEFFECT
% This function can be used to create a chorus audio effect
%
% Input Variables
%   in : single sample of the input signal
%   buffer : used to store delayed samples of the signal
%   n : current sample number used for the LFO
%   depth : range of modulation (milliseconds)
%   rate : speed of modulation (frequency, Hz)
%   predelay : offset of modulation (milliseconds)
%   wet : percent of processed signal (dry = 100 - wet)
%
% See also VIBRATOEFFECT, FLANGEREFFECT

function [out,buffer] = chorusEffect(in,buffer,Fs,n,...
    depth,rate,predelay,wet)

% Calculate time in seconds for the current sample
t = (n-1)/Fs;
lfoMS = depth * sin(2*pi*rate*t) + predelay;
lfoSamples = (lfoMS/1000)*Fs;

% Wet/dry mix
mixPercent = wet;  % 0 = only dry, 100 = only wet
mix = mixPercent/100;


fracDelay = lfoSamples;
intDelay = floor(fracDelay);
```

```
frac = fracDelay - intDelay;

% Store dry and wet signals
drySig = in;
wetSig = (1-frac)*buffer(intDelay,1) + (frac)*buffer(intDelay+1,1);

% Blend parallel paths
out = (1-mix)*drySig + mix*wetSig;

% Linear buffer implemented
buffer = [in ; buffer(1:end-1,1)];
```

Example 15.5: Function: chorusEffect.m

15.5.2 Flanger Effect

Another modulation effect is the **flanger**. It is created using an identical process as the chorus effect, but the delay time is shorter. Rather than the delay range of 10–50 ms of the chorus effect, the flanger effect modulates the delay on a range of 1–50 samples.

Because of the short delay time, audible comb filtering is created due to constructive and destructive interference at different frequencies (Section 12.3.1). While chorus has a "thickening" effect, a flanger has a "thinning" effect due to the notches in the comb filter that reduce the amplitude of many frequencies across the spectrum. As the delay time is slowly modulated between 1, 2, 3, ... samples of delay, the notches in the comb filter sweep across the frequency spectrum.

History

Flanging was first electronically created using analog tape machines. In recording studios, it was common to have multiple tape machines synchronized together in order to increase the number of possible tracks used in a recording (e.g., 24 tracks × 2 machines). In their conventional use, one tape machine would send timecode to another tape machine, such that the second tape machine could automatically synchronize with the first tape machine. If the second tape machine was out of time with the first machine, it would either speed up or slow down until it was synchronized.

To create the flanging effect, an engineer would put duplicate reels of a recording on both machines. This was done to create "parallel" versions of the signal, which could be combined together. During playback and while the machines were synchronized, an engineer would place their finger on the flange of the tape reel in order to temporarily slow down the speed of rotation on one of the machines. This would cause the two machines to no longer be synchronized, with a short time delay between the machines. When the signals from the two machines were combined together, comb filtering was created as the flanger effect.

Flanger Parameters

Similar to the chorus effect, the parameters of *rate*, *depth*, and *pre-delay* are common on the flanger effect. In this case, the delay is shorter and could be given in units of samples instead of milliseconds. Example 15.6 demonstrates an implementation of the flanger effect based on these parameters. In Figures 15.5 and 15.6, a spectrogram is shown of the flanger effect applied to white noise.

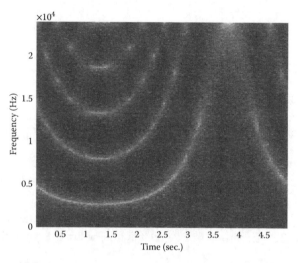

Figure 15.5: Spectrogram of flanger comb filtering on white noise

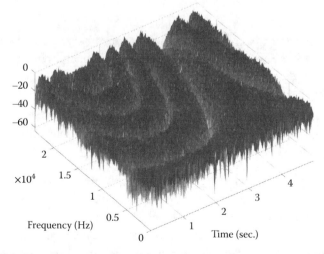

Figure 15.6: Rotated spectrogram (3D) of white noise processed by flanger effect

Examples: Create and run the following m-files in MATLAB.

```matlab
% FLANGEREXAMPLE
% This script creates a flanger effect, applied to white noise.
% Within the processing loop, a feedback flanger can be
% substituted for the feedfoward flanger used by default.
%
% See also FLANGEREFFECT
clear;clc;
Fs = 48000; Ts = 1/Fs;
sec = 5;
lenSamples = sec*Fs;
in = 0.2*randn(lenSamples,1); % White noise input

% Create delay buffer to be hold maximum possible delay time
maxDelay = 50+1;
buffer = zeros(maxDelay,1);

rate = 0.2; % Hz (frequency of LFO)
depth = 4; % Samples (amplitude of LFO)
predelay = 5; % Samples (offset of LFO)
wet = 50; % Wet/dry mix

% Initialize output signal
N = length(in);
out = zeros(N,1);

for n = 1:N

    % Use flangerEffect function
    [out(n,1),buffer] = flangerEffect(in(n,1),buffer,Fs,n,...
        depth,rate,predelay,wet);

    % Use feedbackFlanger function
    %[out(n,1),buffer] = feedbackFlanger(in(n,1),buffer,Fs,n,...
    %    depth,rate,predelay,wet);

end
sound(out,Fs);

% Spectrogram
```

```
nfft = 2048;
window = hann(nfft);
overlap = 128;
[y,f,t,p] = spectrogram(out,window,overlap,nfft,Fs);
surf(t,f,10*log10(p),'EdgeColor','none');
axis xy; axis tight; view(0,90);
xlabel('Time (sec.)');ylabel('Frequency (Hz)');
```

Example 15.6: Script: flangerExample.m

```
% FLANGEREFFECT
% This function can be used to create a flanger audio effect
%
% Input Variables
%   in : single sample of the input signal
%   buffer : used to store delayed samples of the signal
%   n : current sample number used for the LFO
%   depth : range of modulation (samples)
%   rate : speed of modulation (frequency, Hz)
%   predelay : offset of modulation (samples)
%   wet : percent of processed signal (dry = 100 - wet)
%
% See also CHORUSEFFECT, FEEDBACKFLANGER, BARBERPOLEFLANGER

function [out,buffer] = flangerEffect(in,buffer,Fs,n,...
    depth,rate,predelay,wet)

% Calculate time in seconds for the current sample
t = (n-1)/Fs;
lfo = depth * sin(2*pi*rate*t) + predelay;

% Wet/dry mix
mixPercent = wet;  % 0 = only dry, 100 = only wet
mix = mixPercent/100;

fracDelay = lfo;
intDelay = floor(fracDelay);
frac = fracDelay - intDelay;

% Store dry and wet signals
```

```
drySig = in;
wetSig = (1-frac)*buffer(intDelay,1) + (frac)*buffer(intDelay+1,1);


% Blend parallel paths
out = (1-mix)*drySig + mix*wetSig;


% Update buffer
buffer = [in ; buffer(1:end-1,1)];
```

Example 15.7: Function: flangerEffect.m

Feedback Flanger

A different type of flanger effect can be created using feedback delay, instead of feedforward delay. Feedforward delay creates a comb filter with narrow notches to decrease the amplitude at frequencies across the spectrum. The feedback delay creates an inverse-comb filter (Section 13.2.2) with narrow peaks to *increase* the amplitude at frequencies across the spectrum.

To implement the **feedback flanger**, the output signal is stored in the delay buffer instead of the input signal. Otherwise, similar parameters can be used to adjust the characteristics of the LFO and wet/dry blend. A block diagram of a feedback flanger is shown in Figure 15.7. A spectrogram of the effect applied to white noise is shown in Figure 15.8.

Examples: Create the following m-file in MATLAB.

```
% FEEDBACKFLANGER
% This function can be used to create a feedback flanger
%
% Input Variables
%    in : single sample of the input signal
%    buffer : used to store delayed samples of the signal
%    n : current sample number used for the LFO
%    depth : range of modulation (samples)
%    rate : speed of modulation (frequency, Hz)
%    predelay : offset of modulation (samples)
%    wet : percent of processed signal (dry = 100 - wet)
%
% See also FLANGEREFFECT
```

```
function [out,buffer] = feedbackFlanger(in,buffer,Fs,n,...
    depth,rate,predelay,wet)

% Calculate time in seconds for the current sample
t = (n-1)/Fs;
lfo = depth * sin(2*pi*rate*t) + predelay;

% Wet/dry mix
mixPercent = wet;   % 0 = only dry, 100 = only feedback
mix = mixPercent/100;

fracDelay = lfo;
intDelay = floor(fracDelay);
frac = fracDelay - intDelay;

% Store dry and wet signals
drySig = in;
wetSig = (1-frac)*buffer(intDelay,1) + (frac)*buffer(intDelay+1,1);

% Blend parallel paths
out = (1-mix)*drySig + mix*wetSig;

% Feedback is created by storing the output in the
% buffer instead of the input
buffer = [out ; buffer(1:end-1,1)];
```

Example 15.8: Function: `feedbackFlanger.m`

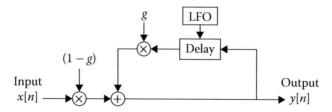

Figure 15.7: Block diagram of feedback flanger

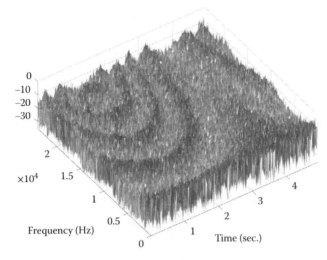

Figure 15.8: Spectrogram of feedback flanger inverted comb filtering

Barber-Pole Flanger

One special type of flanger effect is a **barber-pole flanger**. This effect is perceived as though the sweeping filter is always rising, analogous to visualizing a rotating barber pole. In a conventional flanger, the sweeping filter both rises and falls.

Instead of using a sine-wave LFO, a sawtooth signal is commonly used as the LFO for the barber-pole flanger. Therefore, the delay time for the effect linearly changes over time until the start of the next cycle. The result is a sweeping filter that appears to move in one direction until an abrupt transition to start the next cycle.

In Example 15.10, the barber-pole flanger is implemented by substituting a sawtooth LFO in place of the sine-wave LFO from flanger the effect in Example 15.7.

Examples: Create and run the following m-files in MATLAB.

```
% BARBERPOLEEXAMPLE
%
% This script creates a barber-pole flanger effect using
% a single delay buffer. The delay time is modulated
% by a sawtooth LFO. There is audible distortion each
% time the LFO starts a new cycle.
%
% This script produces a plot with the delay time
```

```matlab
% of the LFO and the spectrogram of white noise processed
% by the effect.
%
% See also BARBERPOLE2EXAMPLE

clear;clc;
Fs = 48000; Ts = 1/Fs;
sec = 8;
lenSamples = sec*Fs;
in = 0.2*randn(lenSamples,1);
t = [0:lenSamples-1].' * (1/Fs);


% Create delay buffer to hold maximum possible delay time
maxDelay = 50+1;
buffer = zeros(maxDelay,1);


rate = 0.5; % Hz (frequency of LFO)
depth = 6; % samples (amplitude of LFO)
predelay = 12; % samples (offset of LFO)


% Wet/dry mix
wet = 50;  % 0 = only dry, 100 = only wet



% Initialize output signal
N = length(in);
out = zeros(N,1);
lfo = zeros(N,1);


for n = 1:N

    % Use flangerEffect.m function
    [out(n,1),buffer,lfo(n,1)] = ...
        barberpoleFlanger(in(n,1),buffer,Fs,n,...
        depth,rate,predelay,wet);

end


sound(out,Fs);
```

```
% Waveform
subplot(3,1,1);
plot(t,lfo);
axis([0 length(t)*(1/Fs) 5 20]);
ylabel('Delay');


% Spectrogram
nfft = 2048; % Length of each time frame
window = hann(nfft); % Calculated windowing function
overlap = 128; % Number of samples for frame overlap
[y,f,tS,p] = spectrogram(out,window,overlap,nfft,Fs);


% Lower subplot
subplot(3,1,2:3);
surf(tS,f,10*log10(p),'EdgeColor','none');
axis xy; axis tight; view(0,90);
xlabel('Time (sec.)');ylabel('Frequency (Hz)');
```

Example 15.9: Script: `barberpoleExample.m`

```
% BARBERPOLEFLANGER
% This function can be used to create a barber-pole flanger
%
% Input Variables
%    in : single sample of the input signal
%    buffer : used to store delayed samples of the signal
%    n : current sample number used for the LFO
%    depth : range of modulation (samples)
%    rate : speed of modulation (frequency, Hz)
%    predelay : offset of modulation (samples)
%    wet : percent of processed signal (dry = 100 - wet)
%
% See also FLANGEREFFECT, FEEDBACKFLANGER, BARBERPOLEFLANGER2


function [out,buffer,lfo] = barberpoleFlanger(in,buffer,Fs,n,...
    depth,rate,predelay,wet)


% Calculate time in seconds for the current sample
t = (n-1)/Fs;
lfo = depth * sawtooth(2*pi*rate*t,0) + predelay;
```

```
% Wet/dry mix
mixPercent = wet;  % 0 = only dry, 100 = only feedback
mix = mixPercent/100;

fracDelay = lfo;
intDelay = floor(fracDelay);
frac = fracDelay - intDelay;

% Store dry and wet signals
drySig = in;
wetSig = (1-frac)*buffer(intDelay,1) + (frac)*buffer(intDelay+1,1);

% Blend parallel paths
out = (1-mix)*drySig + mix*wetSig;

% Feedback is created by storing the output in the
% buffer instead of the input

buffer = [in ; buffer(1:end-1,1)];
```

Example 15.10: Function: `barberpoleFlanger.m`

A basic barber-pole flanger effect has the undersirable characteristic of abrupt changes to the comb filter. This characteristic can be seen in Figure 15.9 by the discontinuity in the spectrogram coinciding with the discontinuity in the sawtooth LFO signal.

To avoid abrupt changes in the flanger's comb filter, a more sophisticated implementation of the barber-pole flanger can be used. A block diagram of the effect is shown in Figure 15.10. In this effect, two modulated delay lines are used in parallel for the purpose of crossfading between them. A sawtooth LFO is used for both delay lines.

A small timing difference is introduced between the LFOs such that the start of each cycle in one LFO overlaps with the end of each cycle in the other LFO. During the overlap, a crossfade is performed between the two delay lines by using the gains, g_1 and g_2. This creates a gradual transition at the start of each cycle of the sawtooth LFO. A MATLAB function called `crossfades` is included with the book's *Additional Content* to synthesize the crossfade gains of the effect. The delay time and amplitude of each path for the barber-pole flanger is shown in Figure 15.11, along with the spectrogram of the effect applied to white noise.

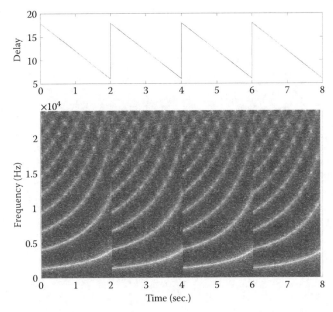

Figure 15.9: Spectrogram of barber-pole flanger with waveform of sawtooth LFO

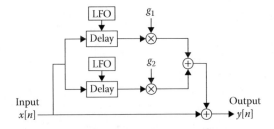

Figure 15.10: Block diagram of crossfade barber-pole flanger

Compared to the barber-pole flanger created without the overlap crossfade shown in Figure 15.9, the spectrogram has a smoother transition at the start of each sawtooth cycle. The crossfade barber-pole flanger is implemented in Example 15.11.

Examples: Create and run the following m-files in MATLAB.

```
% BARBERPOLE2EXAMPLE
% This script creates a barber-pole flanger effect using
% two delay buffers, which gradually fades back and forth
% between the buffers to have a smooth transition at
% the start of the sawtooth ramp.
%
% See also BARBERPOLEEXAMPLE
clear;clc;
```

```
Fs = 48000; Ts = 1/Fs;
sec = 8;
lenSamples = sec*Fs;
in = 0.2*randn(lenSamples,1);
t = [0:lenSamples-1].' * (1/Fs);


% Create delay buffer to be hold maximum possible delay time
maxDelay = 50+1;
buffer = zeros(maxDelay,1);


rate = 0.5; % Hz (frequency of LFO)
depth = 6; % samples (amplitude of LFO)
predelay = 12; % samples (offset of LFO)


% Wet/dry mix
wet = 50;  % 0 = only dry, 100 = only wet



% Initialize output signal
N = length(in);
out = zeros(N,1);
lfo1 = zeros(N,1);
lfo2 = zeros(N,1);
overlap = 18000; % Number of samples of overlap per crossfade
[g1,g2] = crossfades(Fs,lenSamples,rate/2,overlap);


for n = 1:N

    % Use flangerEffect.m function
    [out(n,1),buffer,lfo1(n,1),lfo2(n,1)] = ...
        barberpoleFlanger2(in(n,1),buffer,Fs,n,...
        depth,rate,predelay,wet,g1(n,1),g2(n,1));

end

sound(out,Fs);

% Waveform
subplot(4,1,1);
plot(t,lfo1,t,lfo2);
```

```
axis([0 length(t)*(1/Fs) 8 16]);
ylabel('Delay');


subplot(4,1,2);
plot(t,g1,t,g2); % Crossfade gains
ylabel('Amplitude');


% Spectrogram
nfft = 2048; % Length of each time frame
window = hann(nfft); % Calculated windowing function
overlap = 128; % Number of samples for frame overlap
[y,f,tS,p] = spectrogram(out,window,overlap,nfft,Fs);


% Lower subplot
subplot(4,1,3:4);
surf(tS,f,10*log10(p),'EdgeColor','none');
axis xy; axis tight; view(0,90);
xlabel('Time (sec.)');ylabel('Frequency (Hz)');
```

Example 15.11: Script: `barberpole2Example.m`

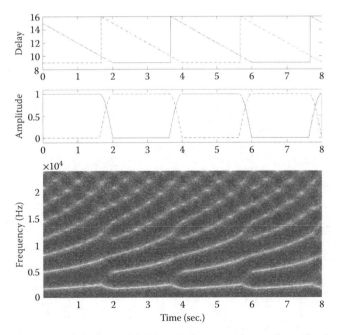

Figure 15.11: Barber-pole flanger: LFO waveform overlap and amplitude crossfade

```
% BARBERPOLEFLANGER2
% This function can be used to create a barber-pole flanger.
% Specifically, the function is meant to crossfade between two
% flangers so that the rising flanger has a smooth transition
% at the start of each sawtooth ramp.
%
% Input Variables
%   in : single sample of the input signal
%   buffer : used to store delayed samples of the signal
%   n : current sample number used for the LFO
%   depth : range of modulation (samples)
%   rate : speed of modulation (frequency, Hz)
%   predelay : offset of modulation (samples)
%   wet : percent of processed signal (dry = 100 - wet)
%   g1 : amplitude of the first flanger in the crossfade
%   g2 : amplitude of the second flanger in the crossfade
%
% See also BARBERPOLEFLANGER

function [out,buffer,lfo1,lfo2] = barberpoleFlanger2(in,buffer,...
    Fs,n,depth,rate,predelay,wet,g1,g2)

% Calculate time in seconds for the current sample
t = (n-1)/Fs;
% Rate/2 because alternating, overlapping LFOs
lfo1 = depth * sawtooth(2*pi*rate/2*t + pi/6,0);
% Hard-clipping at a negative value creates overlap
if lfo1 < -1
    lfo1 = -1;
end
lfo1 = lfo1 + predelay - 2;
lfo2 = depth * sawtooth(2*pi*rate/2*t + 7*pi/6,0);
if lfo2 < -1
    lfo2 = -1;
end

lfo2 = lfo2 + predelay - 2;

% Wet/dry mix
mixPercent = wet;  % 0 = only dry, 100 = only feedback
```

```
mix = mixPercent/100;

fracDelay1 = lfo1;
intDelay1 = floor(fracDelay1);
frac1 = fracDelay1 - intDelay1;

fracDelay2 = lfo2;
intDelay2 = floor(fracDelay2);
frac2 = fracDelay2 - intDelay2;

% Store dry and wet signals
drySig = in;
wetSig = g1 * ((1-frac1)*buffer(intDelay1,1) + ...
    (frac1)*buffer(intDelay1+1,1)) ...
    + g2 * ((1-frac2)*buffer(intDelay2,1) + ...
    (frac2)*buffer(intDelay2+1,1));

% Blend parallel paths
out = (1-mix)*drySig + mix*wetSig;

% Feedback is created by storing the output in the
% buffer instead of the input

buffer = [in ; buffer(1:end-1,1)];
```

Example 15.12: Function: `barberpoleFlanger2.m`

15.6 Pitch Shifter

A consequence of modulating delay time is a frequency change to the input signal. This is an important characteristic in vibrato and chorus. Additionally, this consequence can be used to create the **pitch shifter** effect (Puckette, 2007). It can be implemented using a series delay, as shown in Figure 15.3.

15.6.1 Shifting by Scale Intervals

Audio engineers typically use the pitch shifter effect to change pitch by a constant amount in intervals relative to the musical scale. One example is to increase the pitch of a signal by an octave. In other words, the frequencies of a signal are increased by a factor of two (doubled). Similarly, the pitch shifter could be used to decrease the pitch of signal by an octave. In this case, the frequencies of a signal are decreased by a factor of two (halved). In general, the intervals of the musical scale are **semitones**. Each semitone is a half step in the scale. There are twelve semitones in one octave.

The standard purpose of the pitch shifter is to change the entire signal by the same relative amount. Therefore, all frequencies are increased by a minor third or decreased by a perfect fourth, as examples. Note that this type of pitch shifting is not meant to *tune* the frequencies of a signal to absolute pitches in the music scale (e.g., A440 Hz).

15.6.2 Delay Change and Pitch Change

When delay time increases, or gets longer, over the course of a signal, the period of each cycle gets longer. An increase in the length of a period is related to a decrease in frequency. Conversely, when delay time decreases, or gets shorter, the period of each cycle gets shorter. A decrease in the length of a period is related to an increase in frequency.

Octave Down Example

Consider an example with a modulated delay used to decrease the pitch of a test signal by an octave. Initially, zero samples of delay are used for the first sample. Therefore, the output at the first sample is equal to the first sample of the input. For each subsequent sample, the delay time increases by half a sample. Therefore, the delay time at sample number n is equal to $\frac{n-1}{2}$ samples.

The output at the second sample is the input signal $\frac{1}{2}$ sample in the past. By using a fractional delay, the output at the second sample is interpolated to be halfway between sample 1 and sample 2 of the input. Then, the delay time increases from $\frac{1}{2}$ sample to 1 sample. The output at the third sample is 1 sample in the past from the input, or sample 2. The output at the fourth sample is 1.5 samples in the past from the input, or an interpolated value between samples 2 and 3. The output at the fifth sample is 2 samples in the past from the input, or sample 3.

This process can continue until the final sample of the input signal is used. The effect increases the length of the signal by a factor of 2. Therefore, the frequency is decreased by an octave. In Example 15.13, an input signal of 1 Hz is processed by a pitch shifter. The input signal has a length of one second. The output signal is twice as long and has half the frequency (Figure 15.12).

Examples: Create and run the following m-files in MATLAB.

```
% BASICPITCHDOWN
% This script demonstrates a basic example of
% pitch shifting down an octave created by
% using a modulated time delay.
%
% See also BASICPITCHUP, BASICPITCH
```

```
clc;clear;
% Synthesize 1 Hz test signal
Fs = 48000; Ts = 1/Fs;
t = [0:Ts:1].'; f = 1;
in = sin(2*pi*f*t);
x = [in ; zeros(Fs,1)];  % Zero-pad input because output 2x length


% Initialize loop for pitch decrease
d = 0;       % Initially start with no delay
N = length(x);
y = zeros(N,1);
buffer = zeros(Fs*2,1);


for n = 1:N
    intDelay = floor(d);
    frac = d - intDelay;
    if intDelay == 0   % When there are 0 samples of delay
                       % "y" is based on input "x"
        y(n,1) = (1-frac) * x(n,1) + ...
            frac * buffer(1,1);

    else  % Greater than 0 samples of delay
          % Interpolate between delayed samples "in the past"
        y(n,1) = (1-frac) * buffer(intDelay,1) + ...
            frac * buffer(intDelay+1,1);
    end

    % Store the current input in delay buffer
    buffer = [x(n,1);buffer(1:end-1,1)];

    % Increase the delay time by 0.5 samples
    d = d + 0.5;
end
plot(t,in); hold on;
time = [0:length(y)-1]*Ts; time = time(:);
plot(time,y); hold off;
xlabel('Time (sec.)');ylabel('Amplitude');
legend('Input','Output');
```

Example 15.13: Script: basicPitchDown.m

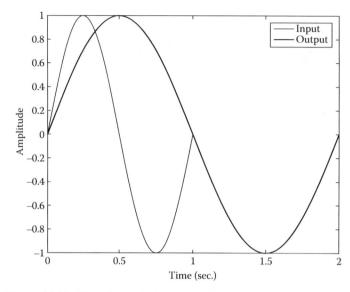

Figure 15.12: Waveform of sine wave pitch shifter one octave down

As a confirmation of the result, consider two points of reference. First, the output at a time of 1 second is based on a delay of $\frac{F_s}{2}$ samples, or 0.5 seconds. Therefore, the output at a time of 1 second is equal to the input at a time of 0.5 seconds. Second, the output at a time of 2 seconds is based on a delay of F_s samples, or 1 second. Therefore, the output at a time of 2 seconds is equal to the input at a time of 1 second. This type of pitch shifting performs **interpolation** to add estimated values in between the samples of the input signal.

Octave Up Example

As a complementary example, consider a pitch shifter effect created to increase the pitch of a signal by an octave. In this case, the output frequency is twice the input frequency. Consequently, the duration of the output signal is half the length of the input signal.

To decrease the period of a signal, a modulated delay time that constantly decreases over the course of a signal is used. An important constraint on the effect is that the delay time cannot be negative. It is necessary for the delay time to be greater than or equal to zero to ensure the effect is only referencing the current sample or samples in the past stored in memory. Therefore, this type of pitch shifter starts with an initial time delay that decreases throughout the signal until the end and finishes with a delay of zero samples.

Consider an example using a signal with a duration of one second. Initially, a delay time equal to the number of samples per second is used, F_s. For each sample, n, the delay time decreases by one sample, $F_s - (n-1)$.

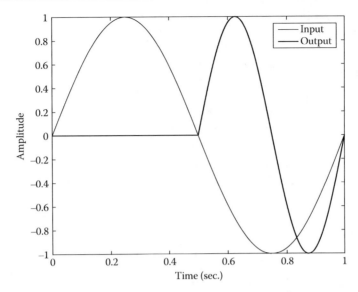

Figure 15.13: Waveform of sine wave pitch shifted one octave up

In Example 15.14, an input signal of 1 Hz is processed by a pitch shifter. The input signal has a length of one second. The period of the output signal is half as long and has twice the frequency, as shown in Figure 15.13.

Examples: Create and run the following m-files in MATLAB.

```
% BASICPITCHUP
% This script demonstrates a basic example of
% pitch shifting up an octave created by
% using a modulated time delay.
%
% See also BASICPITCHDOWN, BASICPITCH
clc;clear;
% Synthesize 1 Hz test signal
Fs = 48000; Ts = 1/Fs;
t = [0:Ts:1].'; f = 1;
x = sin(2*pi*f*t);

d = Fs;                     % Initial delay time

N = length(x);             % Number of samples
y = zeros(N,1);            % Initialize output signal
```

```matlab
buffer = zeros(Fs+1,1);    % Delay buffer
for n = 1:N
    intDelay = floor(d);
    frac = d - intDelay;
    if intDelay == 0
        y(n,1) = (1-frac) * x(n,1) + ...
            frac * buffer(1,1);
    else
        y(n,1) = (1-frac) * buffer(intDelay,1) + ...
            frac * buffer(intDelay+1,1);
    end
    % Store the current input in delay buffer
    buffer = [x(n,1);buffer(1:end-1,1)];

    % Decrease the delay time by 1 sample
    d = d - 1;
end
plot(t,x); hold on;
plot(t,y); hold off;
xlabel('Time (sec.)');ylabel('Amplitude');
legend('Input','Output');
```

Example 15.14: Script: `basicPitchUp.m`

In this example, a sampling rate of 48,000 is used. At a time of 0.5 seconds, or $n = 24{,}001$ samples, the delay time is $48{,}000 - (24{,}001 - 1) = 24{,}000$. Therefore, the output is the input signal at sample number $n - 24{,}000 = 1$, or the first sample of the input. For the next sample of the output, $n = 24{,}002$, the delay time is $48{,}000 - (24{,}002 - 1) = 23{,}999$. Then, the output is the input signal at sample number $n - 23{,}999 = 3$, or the third sample of the input. Each subsequent sample of the output signal is based on the odd sample numbers of the input signal. Therefore, this type of pitch shifting performs **decimation** to remove every other sample of the input signal.

15.6.3 *Rate of Change in Delay Time*

The relative amount of pitch shift is dependent on the *rate of change* for the delay time. For time sample n, the delay time for the current sample, $d[n]$, can be compared to the delay time for the previous sample, $d[n-1]$, to determine the rate of change, $\Delta_d[n]$, using time units of samples. Then the number of semitones, s_t, of pitch shift is determined by the momentary transposition, t_r.

$$\Delta_d[n] = d[n] - d[n-1]$$

$$t_r[n] = 1 - \Delta_d[n]$$

$$s_t[n] = 12 \cdot log_2(t_r[n])$$

To pitch shift a signal by a desired number of semitones, the rate of change for the modulated delay time can be determined by using the following:

$$t_r[n] = 2^{\left(\frac{s_t[n]}{12}\right)}$$

$$\Delta_d[n] = 1 - t_r[n]$$

Octave Examples

Returning to the example for pitch shifting a signal down an octave, the difference of delay time between adjacent samples is $\Delta_d = 0.5$. Therefore, the number of semitones the signal is shifted is $12 \cdot log_2(1 - 0.5) = -12$, or an octave lower.

Similarly, for the example to pitch shift a signal up an octave, the difference of delay time between adjacent samples is $\Delta_d = -1$. In this case, the number of semitones the signal is shifted is $12 \cdot log_2(1 - (-1)) = +12$, or an octave higher.

In Example 15.15, the number of semitones can be set to change the relative amount of pitch shifting. To increase pitch, the number of semitones should be a positive integer. Conversely, using a negative number of semitones decreases pitch.

Examples: Create and run the following m-files in MATLAB.

```
% BASICPITCH
% This script demonstrates a basic example of
% pitch shifting created by
% using a modulated time delay.
clc;clear;
% Synthesize 1 Hz test signal
Fs = 48000; Ts = 1/Fs;
t = [0:Ts:1].';
f = 110;              % Musical note A2 = 110 Hz
in = sin(2*pi*f*t);

% Pitch shift amount
```

```
semitones = -12;      % (-12,-11.....-1,0,1,2,...,11,12,...)
tr = 2^(semitones/12);
dRate = 1 - tr;        % Delay rate of change

% Conditional to handle pitch up or pitch down
if dRate > 0    % Pitch decrease
    d = 0;
    x = [in ; zeros(Fs,1)]; % Prepare for signal to be elongated

else            % Pitch increase
    % Initialize delay so it is always positive
    d = length(in)*-dRate;
    x = in;
end

N = length(x);
y = zeros(N,1);
buffer = zeros(Fs*2,1);
for n = 1:N-1
    intDelay = floor(d);
    frac = d - intDelay;
    if intDelay == 0
        y(n,1) = (1-frac) * x(n,1) + ...
            frac * buffer(1,1);
    else
        y(n,1) = (1-frac) * buffer(intDelay,1) + ...
            frac * buffer(intDelay+1,1);
    end
    % Store the current output in appropriate index
    buffer = [x(n,1);buffer(1:end-1,1)];
    d = d+dRate;
end
plottf(in,Fs);figure; % A2 = 110 Hz
plottf(y,Fs);          % A1 = 55 Hz
```

Example 15.15: Script: basicPitch.m

In Figure 15.14, the musical note A2 = 110 Hz is pitch shifted down 12 semitones to
A1 = 55 Hz.

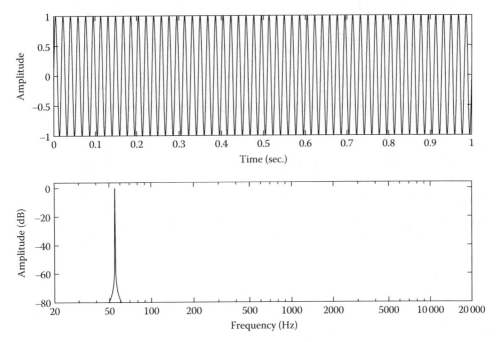

Figure 15.14: A2 = 110 Hz pitch shifted to A1 = 55 Hz

15.6.4 Preserving Signal Duration

When a signal is pitch shifted using a modulated delay, its duration changes. By constantly increasing the delay time, the effect produces a signal longer than the original. By constantly decreasing the delay time, the effect produces a shorter signal.

In some situations for audio engineers, it is acceptable for the processed signal to have different timing than the original signal. In other situations, it would be preferrable for the effect to maintain the same timing after shifting the pitch.

The following is a modification to the pitch shifter effect with the aim to preserve signal duration. Since the amount of pitch shifting is dependent on the rate of change in delay time, not the actual delay time, it is unnecessary for the delay time to increase or decrease indefinitely. Instead, an LFO can be used to modulate the delay time within an acceptable range (e.g., 0–50 ms of delay). In this case, the timing of the processed signal is never different from the input signal by more than the maximum delay time.

To pitch shift by a constant amount, a sawtooth LFO can be used. The slope of the ramp for the sawtooth determines the pitch shifting amount. The LFO is synthesized such that its amplitude never exceeds the maximum delay time. The period, τ, for the sawtooth can be determined by using the maximum delay in samples, $\tau = \frac{maxDelay}{\Delta_d} \cdot T_s$, where Δ_d is the desired rate of change for the delay time and T_s is the sampling period. Then, the LFO can be synthesized using the following:

$$f = \frac{1}{\tau}$$

```
lfo = maxDelay*(0.5*sawtooth(2*pi*f*t)+0.5)
```

In Example 15.16, a pitch shifter effect is implemented using a sawtooth LFO. The delay time is bounded such that it never exceeds a maximum amount of 50 milliseconds. This is done so the output signal has the same duration as the input signal. The pitch shifter effect is applied to an acoustic guitar recording.

Examples: Create and run the following m-files in MATLAB.

```
% LFOPITCH
% This script demonstrates an example of pitch shifting
% using a sawtooth LFO to modulate delay time.
% The result is a signal that has been pitch shifted
% based on the "semitones" variable.
%
% An important aspect of this algorithm is to
% avoid having the processed signal be a different
% length than the original signal. To make this
% possible, the maximum delay time is 50 ms. A
% sawtooth LFO is used to modulate the delay time
% between 0 ms and 50 ms based on the necessary
% rate of change. If the delay time is about to
% go outside of this range, a new cycle of the
% sawtooth begins.
%
% The output signal has audible clicks and pops
% due to the discontinuities of the modulated delay.
% This motivates the use of two parallel delay lines
% that crossfade back and forth to smooth over the
% discontinuities.
%
% See also BASICPITCH, PITCHSHIFTER, PITCHSHIFTEREXAMPLE

clc;clear;close all;
% Synthesize 1 Hz test signal
[in,Fs] = audioread('AcGtr.wav');
Ts = 1/Fs;
```

```matlab
semitones = 1;      % (-12,-11,...,-1,0,1,2,...,11,12,...)
tr = 2^(semitones/12);
dRate = 1 - tr;       % Delay rate of change

maxDelay = Fs * .05;  % Maximum delay is 50 ms

% Conditional to handle pitch up and pitch down
if dRate > 0    % Pitch decrease
    d = 0;

else            % Pitch increase
    % Initialize delay so it is always positive
    d = maxDelay;
end


N = length(in);
out = zeros(N,1);
lfo = zeros(N,1);
buffer = zeros(maxDelay+1,1);
for n = 1:N
    % Determine output of delay buffer
    % which could be a fractional delay time
    intDelay = floor(d);
    frac = d - intDelay;

    if intDelay == 0 % When delay time = zero,
                     % "out" comes "in", not just delay buffer
        out(n,1) = (1-frac) * in(n,1) + ...
            frac * buffer(1,1);
    else
        out(n,1) = (1-frac) * buffer(intDelay,1) + ...
            frac * buffer(intDelay+1,1);
    end

    % Store the current output in appropriate index
    buffer = [in(n,1);buffer(1:end-1,1)];

    % Store the current delay in signal for plotting
    lfo(n,1) = d;
```

```
        d = d+dRate; % Change the delay time for the next loop

        % If necessary, start a new cycle in LFO
        if d < 0
            d = maxDelay;
        elseif d > maxDelay
            d = 0;
        end
    end
    sound(out,Fs);

    t = [0:N-1]*Ts;
    subplot(3,1,1);
    plot(t,lfo); % Crossfade gains
    ylabel('Delay (Samples)'); axis tight;

    % Spectrogram
    nfft = 2048; % Length of each time frame
    window = hann(nfft); % Calculated windowing function
    overlap = 128; % Number of samples for frame overlap
    [y,f,tS,p] = spectrogram(out,window,overlap,nfft,Fs);
    subplot(3,1,2:3);
    surf(tS,f,10*log10(p),'EdgeColor','none');
    axis xy; axis tight; view(0,90);
    xlabel('Time (sec.)');ylabel('Frequency (Hz)');

    % This figure plots the equivalent sawtooth signal
    % to the LFO synthesized during the loop
    figure;
    tau = (maxDelay/dRate) * Ts;
    f = 1/tau;
    plot(t,lfo,t,maxDelay*(0.5*sawtooth(2*pi*f*t)+0.5),'r--');
    axis tight;
```

Example 15.16: Script: lfoPitch.m

The drawback of this approach is that the effect introduces discontinuities in the processed signal. Each time the LFO begins a new cycle, there is a relatively large change in the delay time. At one time sample, n, the delay time may be 0 milliseconds. Then at the next time sample, $n+1$, the delay time may be 50 milliseconds.

The discontinuity in delay time can produce an audible distortion, which sounds like a "click" or "pop." In the spectrogram of the processed signal, the distortion can be seen in the short aperiodic burst of energy across the entire spectrum coinciding with the start of a new cycle in the LFO. In Figure 15.15, the spectrogram of the processed acoustic guitar signal is shown alongside the sawtooth LFO used in the pitch shifter effect.

15.6.5 Parallel Modulated Delay

The pitch shifter effect with a single modulated delay block has a discontinuity problem when a sawtooth LFO is used to control the delay time. To resolve this problem, two parallel delay lines can be used, each with their own sawtooth LFO (Bogdanowicz & Belcher, 1989). A block diagram of this pitch shifter effect is shown in Figure 15.16.

The purpose of the parallel paths is to make it possible to crossfade between the two paths in order to avoid the discontinuity at the start of each LFO cycle. The sawtooth LFOs can be synthesized such that there is an offset in the timing for the start of each cycle. The beginning and end of each cycle overlaps between the two LFOs. A crossfade controlled by the gains, g_1 and g_2, occurs simultaneously with the overlap of the LFOs.

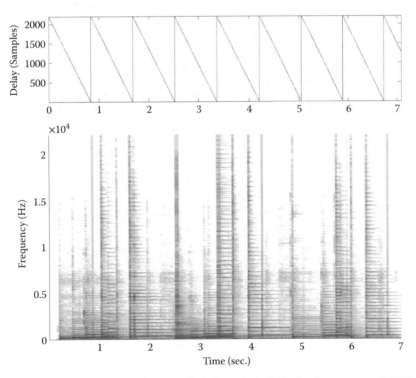

Figure 15.15: Spectrogram of acoustic guitar pitch shifted using a sawtooth LFO

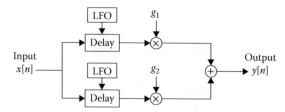

Figure 15.16: Block diagram of pitch shifter effect using parallel delay

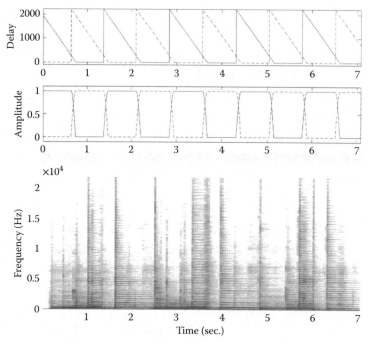

Figure 15.17: Spectrogram of acoustic guitar pitch shifted using parallel delay

Examples of two LFOs, the corresponding crossfade gains, and a spectrogram of a processed acoustic guitar recording are shown in Figure 15.17. Compared to Figure 15.15, the amplitude of the discontinuity distortion has been reduced.

Examples: Create and run the following m-files in MATLAB.

```
% PITCHSHIFTER
% This function implements the pitch shifter
% audio effect by using two parallel delay lines.
% The delay time for each line is modulated by
% a sawtooth LFO. The frequency of the LFO is
% based on the desired number of semitones for
```

```
% the pitch shifter. Both increases and decreases
% in pitch are possible with this function.
%
% Both LFOs repeat a cycle such that the delay
% time stays within a range of 0 ms to 50 ms.
% This way the processed signal is not significantly
% shorter or longer than the original signal.
%
% The cycles of the two LFOs are intentionally
% offset to have an overlap. An amplitude crossfade
% is applied to the delay lines to switch between
% the two during the overlap. The crossfade reduces
% the audibility of the relatively large discontinuity
% in the delay time at the start of each LFO cycle.
%
% See also PITCHSHIFTEREXAMPLE, CROSSFADES, LFOPITCH

function [out] = pitchShifter(in,Fs,semitones)
Ts = 1/Fs;
N = length(in);        % Total number of samples
out = zeros(N,1);
lfo1 = zeros(N,1);     % For visualizing the LFOs
lfo2 = zeros(N,1);


maxDelay = Fs * .05;   % Maximum delay is 50 ms
buffer1 = zeros(maxDelay+1,1);
buffer2 = zeros(maxDelay+1,1);      % Initialize delay buffers


tr = 2^(semitones/12);      % Convert semitones
dRate = 1 - tr;             % Delay rate of change


tau = (maxDelay/abs(dRate))*Ts;  % Period of sawtooth LFO
freq = 1/tau;                    % Frequency of LFO


fade = round((tau*Fs)/8);  % Fade length is 1/8th of a cycle
Hz = (freq/2)*(8/7);       % Frequency of crossfade due to overlap
[g1,g2] = crossfades(Fs,N,Hz,fade);  % Crossfade gains


if dRate > 0    % Pitch decrease
     % Initialize delay so LFO cycles line up with crossfade
```

```
    d1 = dRate * fade;
    d2 = maxDelay;
    d1Temp = d1;  % These variables are used to control
    d2Temp = d2;  % the length of each cycle of the LFO
                  % for the proper amount of overlap

else             % Pitch increase
    % Initialize delay so LFO cycles line up with crossfade
    d1 = maxDelay - maxDelay/8;
    d2 = 0;
    d1Temp = d1;
    d2Temp = d2;
end

% Loop to process input signal
for n = 1:N
    % Parallel delay processing of the input signal
    [out1,buffer1] = fractionalDelay(in(n,1),buffer1,d1);
    [out2,buffer2] = fractionalDelay(in(n,1),buffer2,d2);

    % Use crossfade gains to combine the output of each delay
    out(n,1) = g1(n,1)*out1 + g2(n,1)*out2;

    lfo1(n,1) = d1;  % Save the current delay time
    lfo2(n,1) = d2;  % for plotting

    % The following conditions are set up to control the
    % overlap of the sawtooth LFOs
    if dRate < 0 % Slope of LFO is negative (pitch up)
        d1 = d1 + dRate;
        d1Temp = d1Temp + dRate;
        if d1 < 0
            d1 = 0; % Portion of LFO where delay time = 0
        end
        if d1Temp < -maxDelay * (6/8) % Start next cycle
            d1 = maxDelay;
            d1Temp = maxDelay;
        end

        d2 = d2 + dRate;
```

```
              d2Temp = d2Temp + dRate;
              if d2 < 0
                  d2 = 0; % Portion of LFO where delay time = 0
              end
              if d2Temp < -maxDelay * (6/8) % Start new cycle
                  d2 = maxDelay;
                  d2Temp = maxDelay;
              end

         else  % Slope of LFO is positive (pitch down)

              d1Temp = d1Temp + dRate;
              if d1Temp > maxDelay % Start next cycle
                  d1 = 0;
                  d1Temp = -maxDelay * (6/8);
              elseif d1Temp < 0
                  d1 = 0;        % Portion where delay time = 0
              else
                  d1 = d1 + dRate;
              end


              d2Temp = d2Temp + dRate;
              if d2Temp > maxDelay
                  d2 = 0;
                  d2Temp = -maxDelay * (6/8);
              elseif d2Temp < 0
                  d2 = 0;
              else
                  d2 = d2 + dRate;
              end

         end
     end


% Uncomment for plotting
% t = [0:N-1]*Ts;
% subplot(4,1,1); % Waveform
% plot(t,lfo1,t,lfo2);
```

```
% axis([0 t(end) -100 maxDelay]); ylabel('Delay');
%
% subplot(4,1,2);
% plot(t,g1,t,g2); % Crossfade gains
% axis([0 t(end) -0.1 1.1]);ylabel('Amplitude');
%
%
% nfft = 2048; % Length of each time frame
% window = hann(nfft); % Calculated windowing function
% overlap = 128; % Number of samples for frame overlap
% [y,f,tS,p] = spectrogram(out,window,overlap,nfft,Fs);
%
%
% subplot(4,1,3:4); % Lower subplot spectrogram
% surf(tS,f,10*log10(p),'EdgeColor','none');
% axis xy; axis tight; view(0,90);
% xlabel('Time (sec.)');ylabel('Frequency (Hz)');
```

Example 15.17: Function: `pitchShifter.m`

```
% PITCHSHIFTEREXAMPLE
% This script is an example to demonstrate
% the pitchShifter function for processing
% an audio signal. The desired number of
% semitones can be set from within this script.
%
% See also PITCHSHIFTER, LFOPITCH, BASICPITCH

clc;clear;
% Import acoustic guitar recording for processing
[in,Fs] = audioread('AcGtr.wav');

% Experiment with different values
% -12,-11,....-1,0,1,2,....11,12,...
semitones = 1;

[out] = pitchShifter(in,Fs,semitones);

sound(out,Fs);
% sound(in,Fs); % For comparison
```

Example 15.18: Script: `pitchShifterExample.m`

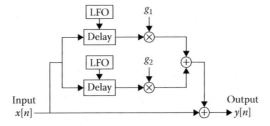

Figure 15.18: Block diagram of harmony pitch shifter effect

15.6.6 Harmony Pitch Shifter Effect

An extension of the pitch shifter is used to create a **harmony effect**. In music, two or more notes with different pitches occurring at the same time produce harmony. This aspect of music can be created through an audio effect by using a pitch shifter in parallel with the unprocessed signal. Then, the pitch of the original signal is supplemented with another pitch from the processed signal.

Since it is necessary in the harmony effect for the timing of the processed signal to be concurrent with the unprocessed signal, the parallel modulated delay can be used as the pitch shifter. A block diagram of the harmony effect is shown in Figure 15.18.

In Example 15.19, a harmony effect is implemented. A pitch shifter is used to decrease the pitch of an acoustic guitar recording by five semitones (i.e., perfect fourth). The processed version of the acoustic guitar is blended with the unprocessed version to create the output signal.

Examples: Create and run the following m-file in MATLAB.

```
% HARMONYEXAMPLE
% This script creates a harmony effect by
% blending together a pitch shifted signal
% with the original, unprocessed signal.
%
% See also PITCHSHIFTER, PITCHSHIFTEREXAMPLE

clc;clear;
% Import acoustic guitar recording for processing
[in,Fs] = audioread('AcGtr.wav');

% Pitch shifted down a perfect fourth
semitones = -5;
processed = pitchShifter(in,Fs,semitones);
```

```
% Blend together input and processed
out = 0.5 * (in + processed);
sound(out,Fs);
% sound(in,Fs); % For comparison
```

Example 15.19: Script: harmonyExample.m

15.7 Modulated All-Pass Filters

All-pass filters (APF) are a special type of system with delay. As discussed in Section 13.3.1, an APF can be created using feedforward and feedback routing with delay. The delay time used with an APF can be modulated to create other time-varying audio effects. Consider the APF shown in Figure 15.19 where the delay time is modulated by a LFO.

On its own, an APF does not change the relative amplitude of different frequencies in the signal. However, the APF does shift the phase of a signal in a frequency dependent way. In Example 15.20, an implemention of an APF is provided using a delay buffer.

Examples: Create and run the following m-file in MATLAB.

```
% ALLPASSFILTER
% This script demonstrates an implementation of an
% all-pass filter using a delay buffer (Direct Form II)

clear;clc;
in = [1; zeros(100,1)];

buffer = zeros(5,1); % Longer buffer than delay length

% Number of samples of delay
delay = 2; % Does not need to be the same length as buffer

g = 0.5;

N = length(in);
out = zeros(N,1);
% Series delay
for n = 1:N

    % Series all-pass filters
```

```
      out(n,1) = g*in(n,1) + buffer(delay,1);
      buffer = [in(n,1) + -g*out(n,1) ; buffer(1:end-1,1)];


  end


  freqz(out);
```

Example 15.20: Script: `allPassFilter.m`

The magnitude and phase response of the APF is shown in Figure 15.20 using an impulse response of the system. The magnitude response is flat for all frequencies. The phase response shows a phase shift introduced by the APF across the spectrum. Initially for low frequencies, the phase shift is near 0°. The phase shift gradually increases with increasing frequency, such that a phase shift of 180° occurs at half of the Nyquist frequency. This transition continues until the Nyquist frequency where a phase shift of 360° occurs.

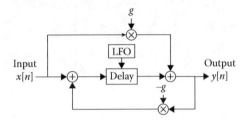

Figure 15.19: Block diagram of APF with a modulated delay

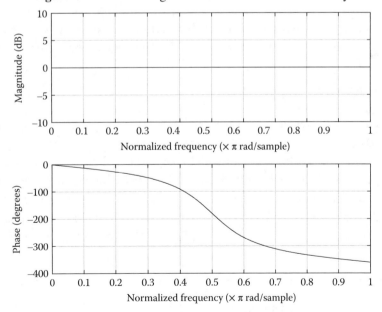

Figure 15.20: Frequency response of an APF

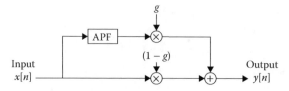

Figure 15.21: Block diagram for phaser effect

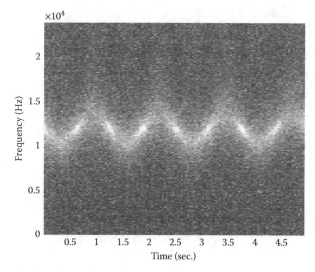

Figure 15.22: Spectrogram of white noise processed by a phaser effect

Modulating the delay time changes the phase shift across the spectrum. As a important point of reference, the frequency where a phase shift of 180° occurs can be increased or decreased.

15.7.1 Phaser Effect

The **phaser** audio effect is created by modulating the delay time of an APF. This effect uses constructive and destructive interference to decrease the amplitude of a narrow range of frequencies. By using the APF in parallel with the dry signal, destructive interference occurs for frequencies near where the APF introduces a 180° phase shift. Furthermore, constructive interference occurs for frequencies near where the APF introduces a phase shift of 0° or 360°. A block diagram for the phaser effect is shown in Figure 15.21 and implemented in Example 15.21. A spectrogram of the effect applied to white noise is shown in Figure 15.22.

Examples: Create and run the following m-files in MATLAB.

```
% PHASEREXAMPLE
% This script demonstrates the use of a phaser
% function to add the effect to white noise. Parameters
```

```
% of the phaser effect include the rate and depth of
% the LFO. In this implementation, the delay time of
% an APF (Direct Form II) is modulated.
%
% See also PHASEREFFECT

clear;clc;
Fs = 48000; Ts = 1/Fs;
sec = 5;
lenSamples = sec*Fs;
in = 0.2*randn(lenSamples,1);


rate = 0.8; % Hz (frequency of LFO)
depth = 0.3; % samples (amplitude of LFO)

% Initialize delay buffers
buffer = zeros(3,1); % All-pass filter

% Wet/dry mix
wet = 50;

% Initialize output signal
out = zeros(size(in));

for n = 1:length(in)

    % Use phaser effect function
    [out(n,1),buffer] = phaserEffect(in(n,1),buffer,Fs,n,...
    depth,rate,wet);

end

sound(out,Fs);

% Spectrogram
nfft = 2048;
window = hann(nfft);
overlap = 128;
[y,f,t,p] = spectrogram(out,window,overlap,nfft,Fs);
surf(t,f,10*log10(p),'EdgeColor','none');
```

```
axis xy; axis tight; view(0,90); %view(-30,60);
xlabel('Time (sec.)');ylabel('Frequency (Hz)');
```

Example 15.21: Script: `phaserExample.m`

```
% PHASEREFFECT
% This function can be used to create a phaser audio effect
%
% Input Variables
%   in : single sample of the input signal
%   buffer : used to store delayed samples of the signal
%   n : current sample number used for the LFO
%   depth : range of modulation (samples)
%   rate : speed of modulation (frequency, Hz)
%   wet : percent of processed signal (dry = 100 - wet)
%
% See also BIQUADPHASER

function [out,buffer] = phaserEffect(in,buffer,Fs,n,...
    depth,rate,wet)


% Calculate time in seconds for the current sample
t = (n-1)/Fs;
lfo = depth * sin(2*pi*rate*t)+2;

% Wet/dry mix
mixPercent = wet;  % 0 = only dry, 100 = only wet
mix = mixPercent/100;


fracDelay = lfo;
intDelay = floor(fracDelay);
frac = fracDelay - intDelay;

% Store dry and wet signals
drySig = in;


g = 0.25;
% All-pass filter
wetSig = g*in + ((1-frac)*buffer(intDelay,1) + ...
    frac*buffer(intDelay+1,1));
```

```
% Blend parallel paths
out = (1-mix)*drySig + mix*wetSig;


buffer = [in + -g*wetSig ; buffer(1:end-1,1)];
```

Example 15.22: Function: `phaserEffect.m`

Bi-Quadratic Phaser Effect

Another approach for creating the phaser effect uses a bi-quad filter (Section 13.3.2) to implement the all-pass filter on the parallel path shown in Figure 15.21. With the bi-quadratic filter, the frequency where a phase shift of 180° occurs can be chosen to be any frequency across the spectrum. This allows for additional control of the effect's processing.

In this implementation, a conventional fractional delay is not used. Instead, the bi-quad coefficients, $\{b_0, b_1, b_2, a_0, a_1, a_2\}$, are changed. As part of the effect, an LFO is synthesized to provide the frequency at which the phase shift of 180° occurs. The value of the LFO at a time sample can be used to determine the bi-quad coefficients based on Table 13.1. The width of the phaser's notch can be controlled using the parameter Q for the bi-quad filter. In Example 15.23, a phaser effect is implemented using a bi-quad APF. A spectrogram of the effect applied to white noise is shown in Figure 15.23.

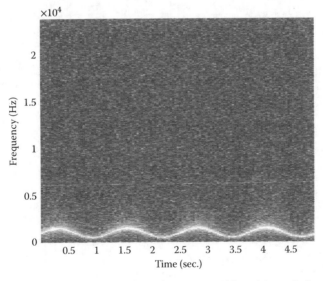

Figure 15.23: Spectrogram of white noise processed by a bi-quad phaser effect

Examples: Create and run the following m-files in MATLAB.

```matlab
% PHASEREXAMPLE2
% This script implements a phaser effect using
% a bi-quad filter as the APF.
%
% See also BIQUADPHASER

clear;clc;
Fs = 48000; Ts = 1/Fs;
sec = 5;
lenSamples = sec*Fs;
in = 0.2*randn(lenSamples,1);

rate = 0.8; % Hz (frequency of LFO)
centerFreq = 1000; % Hz (center freq of LFO)
depth = 500; % Hz (LFO range = 500 to 1500)

% Feedforward delay buffer
ff = [0 ; 0]; % ff(n,1) = n samples of delay

% Feedback delay buffer
fb = [0 ; 0]; % fb(n,1) = n samples of delay

% Bandwidth of phaser (wide or narrow notch)
Q = 0.5;

% Wet/dry mix
wet = 50;

% Initialize output signal
out = zeros(size(in));

for n = 1:length(in)
    t = (n-1) * Ts;
    lfo = depth*sin(2*pi*rate*t) + centerFreq;

    % Use Bi-quad phaser effect function
    [out(n,1),ff,fb] = biquadPhaser(in(n,1),Fs,...
        lfo,Q,ff,fb,wet);
```

```
end


sound(out,Fs);

% Spectrogram
nfft = 2048;
window = hann(nfft);
overlap = 128;
[y,f,t,p] = spectrogram(out,window,overlap,nfft,Fs);
surf(t,f,10*log10(p),'EdgeColor','none');
axis xy; axis tight; view(0,90); %view(-30,60);
xlabel('Time (sec.)');ylabel('Frequency (Hz)');
```

Example 15.23: Script: phaserExample2.m

```
% BIQUADPHASER
% This function can be used to create a phaser audio effect
% by using a bi-quad APF
%
% Input Variables
%   in : single sample of the input signal
%   Fs : sampling rate
%   lfo : used to determine the frequency of APF
%   ff : buffer for feedfoward delay
%   fb : buffer for feedback delay
%   wet : percent of processed signal (dry = 100 - wet)
%
% Use Table 13.1 to calculate APF bi-quad coefficients
%
% See also PHASEREFFECT, BIQUADWAH

function [out,ff,fb] = biquadPhaser(in,Fs,...
        lfo,Q,ff,fb,wet)

% Convert value of LFO to normalized frequency
w0 = 2*pi*lfo/Fs;
% Normalize bandwidth
alpha = sin(w0)/(2*Q);
```

```
b0 = 1-alpha;    a0 = 1+alpha;
b1 = -2*cos(w0); a1 = -2*cos(w0);
b2 = 1+alpha;    a2 = 1-alpha;

% Wet/dry mix
mixPercent = wet;  % 0 = only dry, 100 = only wet
mix = mixPercent/100;

% Store dry and wet signals
drySig = in;

% All-pass filter
wetSig = (b0/a0)*in + (b1/a0)*ff(1,1) + ...
    (b2/a0)*ff(2,1) - (a1/a0)*fb(1,1) - (a2/a0)*fb(2,1);

% Blend parallel paths
out = (1-mix)*drySig + mix*wetSig;

% Iterate buffers for next sample
ff(2,1) = ff(1,1);
ff(1,1) = in;
fb(2,1) = fb(1,1);
fb(1,1) = wetSig;
```

Example 15.24: Function: `biquadPhaser.m`

15.8 Modulated Low-Pass Filters

As shown in Section 15.7.1, a bi-quad filter can be used to create modulated filter effects. By changing the filter type from an APF (phaser) to an LPF, an auto-wah effect can be created.

15.8.1 Auto-Wah Effect

An auto-wah (automatic-wah) effect is based on a resonant LPF. As an input signal is filtered over time, the cutoff frequency is modulated by a LFO. Due to the resonance of the LPF, the effect gives the impression of a vocalized "wah" utterance applied to the input signal. In Example 15.25, an auto-wah effect is demonstrated using a bi-quad filter.

Examples: Create and run the following m-files in MATLAB.

```matlab
% AUTOWAHEXAMPLE
% This script implements an auto-wah effect using
% a bi-quad filter as the resonant LPF.
%
% See also BIQUADWAH


clear;clc;
[in,Fs] = audioread('AcGtr.wav');
Ts = 1/Fs;
rate = 0.5; % Hz (frequency of LFO)
centerFreq = 1500; % Hz (center freq of LFO)
depth = 1000; % Hz (LFO range = 500 to 2500)

% Feedforward delay buffer
ff = [0 ; 0]; % ff(n,1) = n samples of delay

% Feed-back Delay Buffer
fb = [0 ; 0]; % fb(n,1) = n samples of delay

% Bandwidth of resonant LPF
Q = 7;

% Wet/dry mix
wet = 100;

% Initialize output signal
out = zeros(size(in));

for n = 1:length(in)
    t = (n-1) * Ts;
    lfo = depth*sin(2*pi*rate*t) + centerFreq;

    % Use bi-quad wah effect function
    [out(n,1),ff,fb] = biquadWah(in(n,1),Fs,...
        lfo,Q,ff,fb,wet);

end
```

```
sound(out,Fs);

% Spectrogram - notice resonances in shape of sine wave
nfft = 2048;
window = hann(nfft);
overlap = 128;
[y,f,t,p] = spectrogram(out,window,overlap,nfft,Fs);
surf(t,f,10*log10(p),'EdgeColor','none');
axis xy; axis tight; view(0,90);
xlabel('Time (sec.)');ylabel('Frequency (Hz)');
```

Example 15.25: Script: `autoWahExample.m`

```
% BIQUADWAH
% This function can be used to create a wah-wah audio effect
%
% Input Variables
%   in : single sample of the input signal
%   Fs : sampling rate
%   lfo : used to determine the frequency of LPF
%   ff : buffer for feedfoward delay
%   fb : buffer for feedback delay
%   wet : percent of processed signal (dry = 100 - wet)
%
% Use Table 13.1 to calculate LPF bi-quad coefficients
%
% See all BIQUADPHASER

function [out,ff,fb] = biquadWah(in,Fs,...
        lfo,Q,ff,fb,wet)




% Convert value of LFO to normalized frequency
w0 = 2*pi*lfo/Fs;
% Normalize bandwidth
alpha = sin(w0)/(2*Q);

b0 = (1-cos(w0))/2;    a0 = 1+alpha;
b1 = 1-cos(w0);        a1 = -2*cos(w0);
```

```
b2 = (1-cos(w0))/2;     a2 = 1-alpha;


% Wet/dry mix
mixPercent = wet;  % 0 = only dry, 100 = only wet
mix = mixPercent/100;


% Store dry and wet signals
drySig = in;


% Low-pass filter
wetSig = (b0/a0)*in + (b1/a0)*ff(1,1) + ...
    (b2/a0)*ff(2,1) - (a1/a0)*fb(1,1) - (a2/a0)*fb(2,1);


% Blend parallel paths
out = (1-mix)*drySig + mix*wetSig;


% Iterate buffers for next sample
ff(2,1) = ff(1,1);
ff(1,1) = in;
fb(2,1) = fb(1,1);
fb(1,1) = wetSig;
```

Example 15.26: Function: `biquadWah.m`

Bibliography

Bogdanowicz, K., & Belcher, R (1989). Using multiple processors for real-time audio effects. *Audio Engineering Society Conference*, 7.

Dutilleux, P., & Zölzer, U. (2002) Delays. In U. Zölzer (Ed.), *DAFX: Digital audio effects* (p. 68), New York, NY: John Wiley & Sons.

Puckette, M. (2007) *Theory and technique of electronic music* (p. 202). Singapore: World Scientific Publishing Co. Pte. Ltd.

Chapter 16

Algorithmic Reverb Effects
Schroeder, Moorer, and FDN

16.1 Introduction: Algorithmic Reverb Effects

This chapter introduces several different algorithms for creating artificial, digital reverberation (reverb). These effects create many delayed repetitions of an input signal to simulate reflections in an acoustic space. The possible algorithms for creating artificial reverb are only limited by an engineer's creativity. This chapter presents three algorithms that can serve as foundational approaches: Schroeder, Moorer, and Jot feedback delay network (FDN).

Rather than creating artificial reverberation by using convolution with a measured impulse response (Section 11.8.3), **algorithmic reverb** is based more on the concept of creating successively repeating delays using feedback. Therefore, algorithmic reverb may not be as accurate as a measured impulse response for re-creating a physical, acoustic space. However, convincing results can still be achieved with a reduced computational cost.

16.1.1 Combining Modulated Delay Effects

A common processing block used in algorithmic reverb is the modulated delay. Several types of modulated delay are discussed in Chapter 15 including series, parallel, feedforward, feedback, and the all-pass filter. These types of modulated delay can be combined together in creative ways to simulate reverberation in each algorithm.

16.1.2 Advantages of Algorithmic Reverb

Compared to convolution reverb (Section 11.8.3), there are advantages to using algorithmic reverb. For some styles of music, the sound of algorithmic reverb has become an essential part of the genre's asthetic. Many classic recordings from the 1970s, 1980s, and 1990s used algorithmic reverb to give the music a perception of space.

When algorithmic reverbs were originally developed, computers were not able to perform convolution fast enough to make it practical in many situations. This motivated the creation of computationally efficient approaches for simulating reverb.

Memory and CPU Calculations

To perform convolution with an impulse response of a measured reverb tail of several seconds, a computer must store in memory 100,000+ samples of the impulse response and also the input signal. Additionally, the computer must perform 100,000+ multiplication and addition operations for each sample in the signal. This is due to the underlying feedforward approach for creating delay.

In algorithmic reverb, the use of feedback makes it possible to create the effect by storing fewer samples in memory ($\sim$1,000–10,000+) and performing fewer multiplication and addition operations per sample ($\sim$10–100+).

16.2 Schroeder

One of the earliest published algorithms for creating reverb is Schroeder (1962). This initial design is based on series and parallel combinations of feedback comb filters and all-pass filters. Each block of the algorithm creates repetitions of an input signal. As these repetitions are sent through subsequent processing blocks, additional repetitions are generated to build up the echo density of the reverb. Bear in mind that the feedback comb filters and the all-pass filters are infinite impulse response filters.

16.2.1 Feedback Comb Filters

An important block used in the Schroeder reverb is the feedback comb filter (FBCF), which was discussed in Chapter 13. Because the FBCF is used on the overall "wet" path of the reverb, it is not necessary for each block to blend the delayed signal with the unprocessed input signal. Rather, a modified version of the FBCF can be used such that the output of the block is only the delayed signal, while feedback sends the output of the delay buffer back to the input of the delay buffer. An example of the FBCF for the Schroeder reverb is shown in Figure 16.1.

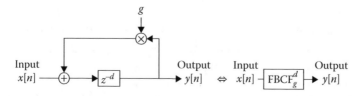

Figure 16.1: Block diagram of feedback comb filter for Schroeder reverb

An implementation of an FBCF without modulated delay is shown in Example 16.1. A test script for the function is demonstrated in Example 16.2.

Examples: Create the following function in MATLAB.

```
% FBCFNOMOD
% This function creates a feedback comb filter by
% processing an individual input sample and updating
% a delay buffer used in a loop to index each sample
% in a signal. This implementation does not use
% fractional delay. Therefore, the delay time
% cannot be modulated.
%
% Input Variables
%   n : current sample number of the input signal
%   delay : samples of delay
%   fbGain : feedback gain (linear scale)
%
% See also FBCF

function [out,buffer] = fbcfNoMod(in,buffer,n,delay,fbGain)

% Determine indexes for circular buffer
len = length(buffer);
indexC = mod(n-1,len) + 1; % Current index
indexD = mod(n-delay-1,len) + 1; % Delay index

out = buffer(indexD,1);

% Store the current output in appropriate index
buffer(indexC,1) = in + fbGain*buffer(indexD,1);
```

Example 16.1: Function: `fbcfNoMod.m`

Modulated Delay for Ringing

The use of feedback to create repetitions also causes a frequency-dependent change in amplitude. Feedback delay creates an inverse comb filter with spectral peaks at different frequencies (Section 13.2.2). The amplitude at some frequencies builds up and resonates, while other frequencies do not. The resonant frequencies are directly related to the delay time (Section 12.3.1).

By constantly changing the delay time during the effect, it is possible to prevent the amplitude from building up at any one frequency. Therefore, delay time modulation has the effect of decreasing the spectral resonances due to feedback. Perceptually, this makes the reverb sound more natural by decreasing ringing.

Modulating delay by as little ±0.1 samples using an LFO can improve perceived performance. Typically, a delay time modulation is selected in the range of ±2 to ±8 samples. Some algorithms have used delay time modulation as high as ±16 samples. As delay time modulation increases, listeners perceive a chorus effect (Section 15.5.1) in addition to the reverb.

Examples: Create and run the following m-files in MATLAB.

```
% FBCFEXAMPLE
% This script uses a feedback comb filter (FBCF)
% function, applied to an acoustic guitar recording.
%
% See also FBCFNOMOD, FBCF
clear;clc;

[in,Fs] = audioread('AcGtr.wav');

maxDelay = ceil(.05*Fs);  % maximum delay of 50 ms
buffer = zeros(maxDelay,1);    % initialize delay buffer

d = .04*Fs; % 40 ms of delay
g = -0.7;   % feedback gain value

rate = 0.6; % Hz (frequency of LFO)
amp = 6; % Range of +- 6 samples for delay

% Initialize output signal
N = length(in);
out = zeros(N,1);

for n = 1:N

    % Uncomment to use fbcfNoMod.m function
    %[out(n,1),buffer] = fbcfNoMod(in(n,1),buffer,n,d,g);
```

```
    % Use fbcf.m function
    [out(n,1),buffer] = fbcf(in(n,1),buffer,Fs,n,...
        d,g,amp,rate);


end


sound(out,Fs);
```

Example 16.2: Script: fbcfExample.m

```
% FBCF
% This function creates a feedback comb filter by
% processing an individual input sample and updating
% a delay buffer used in a loop to index each sample
% in a signal. Fractional delay is implemented to make
% it possible to modulate the delay time.
%
% Input Variables
%   n : current sample number of the input signal
%   delay : samples of delay
%   fbGain : feedback gain (linear scale)
%   amp : amplitude of LFO modulation
%   rate : frequency of LFO modulation
%
% See also FBCFNOMOD

function [out,buffer] = fbcf(in,buffer,Fs,n,delay,fbGain,amp,rate)

% Calculate time in seconds for the current sample
t = (n-1)/Fs;
fracDelay = amp * sin(2*pi*rate*t);
intDelay = floor(fracDelay);
frac = fracDelay - intDelay;

% Determine indexes for circular buffer
len = length(buffer);
indexC = mod(n-1,len) + 1; % Current index
indexD = mod(n-delay-1+intDelay,len) + 1; % Delay index
```

```
indexF = mod(n-delay-1+intDelay+1,len) + 1; % Fractional index

out = (1-frac)*buffer(indexD,1) + (frac)*buffer(indexF,1);

% Store the current output in appropriate index
buffer(indexC,1) = in + fbGain*out;
```

Example 16.3: Function: `fbcf.m`

Series

There are different ways to combine the processing blocks of algorithmic reverb. One way is to combine the processing blocks using a series connection. In this case, the input signal is used as the input to the first processing block. The output of the first processing block is used as the input to the second processing block. A series connection of FBCFs is shown in Figure 16.2.

Examples: Create the following m-file in MATLAB.

```
% FBCFSERIESEXAMPLE
% This script uses series feedback comb filter (FBCF)
% functions, applied to an acoustic guitar recording.
%
% See also FBCFPARALLELEXAMPLE
clear;clc;

[in,Fs] = audioread('AcGtr.wav');

maxDelay = ceil(.07*Fs);  % max delay of 70 ms
buffer1 = zeros(maxDelay,1);
buffer2 = zeros(maxDelay,1);

d1 = fix(.042*Fs); % 42 ms of delay
g1 = 0.5;
d2 = fix(.053*Fs); % 53 ms of delay
g2 = -0.5;

rate1 = 0.6; % Hz (frequency of LFO)
amp1 = 6; % Range of +- 6 samples for delay
rate2 = 0.5; % Hz (frequency of LFO)
amp2 = 8; % Range of +- 8 samples for delay
```

```
% Initialize output signal
N = length(in);
out = zeros(N,1);


for n = 1:N

    % Two series FBCFs
    [w,buffer1] = fbcf(in(n,1),buffer1,Fs,n,...
        d1,g1,amp1,rate1);

    % The output "w" of the first FBCF is used
    % as the input to the second FBCF
    [out(n,1),buffer2] = fbcf(w,buffer2,Fs,n,...
        d2,g2,amp2,rate2);

end

sound(out,Fs);
```

Example 16.4: Script: `fbcfSeriesExample.m`

Parallel

Another way to combine the processing blocks is a parallel connection. In this case, the input signal is used as the input to both of the processing blocks. The output of each processing block is added together to create the output signal. A parallel connection of FBCFs is shown in Figure 16.3.

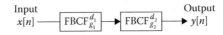

Figure 16.2: Block diagram of series feedback comb filters

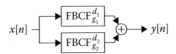

Figure 16.3: Block diagram of parallel feedback comb filters

Examples: Create the following m-file in MATLAB.

```matlab
% FBCFPARALLELEXAMPLE
% This script uses parallel feedback comb filter (FBCF)
% functions, applied to an acoustic guitar recording.
%
% See also FBCFSERIESEXAMPLE
clear;clc;

[in,Fs] = audioread('AcGtr.wav');

maxDelay = ceil(.07*Fs);   % max delay of 70 ms
buffer1 = zeros(maxDelay,1);
buffer2 = zeros(maxDelay,1);

d1 = fix(.047*Fs); % 47 ms of delay
g1 = 0.5;
d2 = fix(.053*Fs); % 53 ms of delay
g2 = -0.5;

rate1 = 0.6; % Hz (frequency of LFO)
amp1 = 6; % Range of +- 6 samples for delay
rate2 = 0.5; % Hz (frequency of LFO)
amp2 = 8; % Range of +- 8 samples for delay

% Initialize output signal
N = length(in);
out = zeros(N,1);

for n = 1:N

    % Two parallel FBCFs
    [w1,buffer1] = fbcf(in(n,1),buffer1,Fs,n,...
        d1,g1,amp1,rate1);
    % Both FBCFs receive "in" to create
    % parallel processing
    [w2,buffer2] = fbcf(in(n,1),buffer2,Fs,n,...
        d2,g2,amp2,rate2);
    % The output of each FBCF is summed together
```

```
     % to complete parallel processing
     out(n,1) = w1 + w2;
 end


 sound(out,Fs);
```

Example 16.5: Script: fbcfParallelExample.m

Reverb Time from Feedback Amplitude

For the Schroeder reverb, the primary method for an audio engineer to control reverb time is to change the gain on the feedback path of the FBCF. Increasing the gain lengthens the reverb time because the repetitions persist in the feedback system for a longer time. Decreasing the gain shortens the reverb time because the repetitions decrease in amplitude more quickly.

16.2.2 All-Pass Filter: Colorless Delays

In addition to the FBCF, the other important processing block of the Schroeder reverb is the all-pass filter (APF). The implementation of this APF is similar to the design described in Section 13.7.1 and used in the phaser effect in Section 15.7.1.

Instead of using one or two samples of the delay, the APFs of the Schroeder reverb use a delay of a relatively longer time. Therefore, the APFs create audible repetitions for the reverb effect and should not merely be considered as shifting a signal's phase.

Compared to the FBCF, there is a benefit to incorporating APFs in a reverb algorithm. One undesirable consequence of the FBCF is the peak resonances occurring across the spectrum, introducing unnatural ringing to the reverb sound. In theory, the use of APFs can accomplish spectrally flat, colorless, repetitions. In practice, APFs with a relatively long delay time can still exibit temporary resonant qualities, although less pronounced than a FBCF. A example of the block diagram for an APF with arbitrary delay length is shown in Figure 16.4.

Examples: Create and run the following m-files in MATLAB.

```
% APFEXAMPLE
% This script uses an all-pass filter
% function, applied to an acoustic guitar recording.
%
% See also APF
```

```
clear;clc;

[in,Fs] = audioread('AcGtr.wav');

maxDelay = max(ceil(.05*Fs));  % maximum delay of 50 ms
buffer = zeros(maxDelay,1);

d = ceil(.042*Fs); % 42 ms of delay
g = 0.9;

rate = 0.9; % Hz (frequency of LFO)
amp = 6; % Range of +- 6 samples for delay

% Initialize output signal
N = length(in);
out = zeros(N,1);

for n = 1:N

    % Use apf.m function
    [out(n,1),buffer] = apf(in(n,1),buffer,Fs,n,...
        d,g,amp,rate);

end

sound(out,Fs);
```

Example 16.6: Script: `apfExample.m`

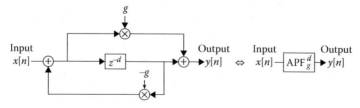

Figure 16.4: Block diagram of all-pass filter for Schroeder reverb

```
% APF
% This function creates an all-pass filter by
% processing an individual input sample and updating
% a delay buffer used in a loop to index each sample
% in a signal.
%
% Input Variables
%   n : current sample number of the input signal
%   delay : samples of delay
%   gain : feedback gain (linear scale)
%   amp : amplitude of LFO modulation
%   rate : frequency of LFO modulation

function [out,buffer] = apf(in,buffer,Fs,n,delay,gain,amp,rate)

% Calculate time in seconds for the current sample
t = (n-1)/Fs;
fracDelay = amp * sin(2*pi*rate*t);
intDelay = floor(fracDelay);
frac = fracDelay - intDelay;

% Determine indexes for circular buffer
len = length(buffer);
indexC = mod(n-1,len) + 1; % Current index
indexD = mod(n-delay-1+intDelay,len) + 1; % Delay index
indexF = mod(n-delay-1+intDelay+1,len) + 1; % Fractional index

% Temp variable for output of delay buffer
w = (1-frac)*buffer(indexD,1) + (frac)*buffer(indexF,1);

% Temp variable used for the node after the input sum
v = in + (-gain*w);

% Summation at output
out = (gain * v) + w;

% Store the current input to delay buffer
buffer(indexC,1) = v;
```

Example 16.7: Function: apf.m

Diffusion from APF Gain

For the Schroeder reverb (as well as several other algorithms), the feedforward and feedback gain in the APF can be used to control diffusion. By increasing both of the gains, an increase in the density of repetitions is achieved. This simulates the scattering of echoes occurring in highly diffuse environments. By decreasing the gains, the density of repetitions is decreased.

16.2.3 Schroeder Algorithm Implementation

The implementation of the Schroeder algorithm involves routing the input signal through four (or more) parallel feedback comb filters. Then, the output of each filter is summed together. Finally, the summed signal is sent through two (or more) series all-pass filters to create the output signal. This process represents the processed (wet) path for the reverb to be blended with an unprocessed (dry) path. The delay times used for one version of the Schroder algorithm are shown in Table 16.1. A block diagram based on these delay times is shown in Figure 16.5.

The delay times are modulated during implementation of the algorithm. By selecting different delay times, a reverb with an alternative sound can be created. By shortening the delay times of the FBCFs, a smaller room can be simulated. By lengthening the delay times of the FBCFs,

Table 16.1: Delay times in Schroeder reverb

Milliseconds	Samples of Delay (48 kHz)
43.7	2098
41.1	1973
37.1	1781
29.7	1426
5.0	240
1.7	82

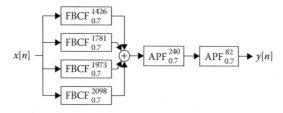

Figure 16.5: Block diagram of Schroeder reverb

a larger room can be simulated. By changing the gain of the FBCFs, the reverb time can be increased ($g > 0.7$) or decreased ($g < 0.7$).

Examples: Create the following m-file in MATLAB.

```
% SCHROEDERREVERB
% This script implements the Schroeder reverb algorithm
% by using feedback comb filters (fbcf) and all-passs
% filters (apf)
%
% See also FBCF, APF
clear;clc;

[in,Fs] = audioread('AcGtr.wav');

% Max delay of 70 ms
maxDelay = ceil(.07*Fs);
% Initialize all buffers (there are 6 total - 4 FBCF, 2 APF)
buffer1 = zeros(maxDelay,1); buffer2 = zeros(maxDelay,1);
buffer3 = zeros(maxDelay,1); buffer4 = zeros(maxDelay,1);
buffer5 = zeros(maxDelay,1); buffer6 = zeros(maxDelay,1);

% Delay and gain parameters
d1 = fix(.0297*Fs); g1 = 0.75;
d2 = fix(.0371*Fs); g2 = -0.75;
d3 = fix(.0411*Fs); g3 = 0.75;
d4 = fix(.0437*Fs); g4 = -0.75;
d5 = fix(.005*Fs); g5 = 0.7;
d6 = fix(.0017*Fs); g6 = 0.7;

% LFO parameters
rate1 = 0.6; amp1 = 8;
rate2 = 0.71; amp2 = 8;
rate3 = 0.83; amp3 = 8;
rate4 = 0.95; amp4 = 8;
rate5 = 1.07; amp5 = 8;
rate6 = 1.19; amp6 = 8;

% Initialize output signal
N = length(in);
```

```
out = zeros(N,1);

for n = 1:N

    % Four parallel FBCFs
    [w1,buffer1] = fbcf(in(n,1),buffer1,Fs,n,...
        d1,g1,amp1,rate1);

    [w2,buffer2] = fbcf(in(n,1),buffer2,Fs,n,...
        d2,g2,amp2,rate2);

    [w3,buffer3] = fbcf(in(n,1),buffer3,Fs,n,...
        d3,g3,amp3,rate3);

    [w4,buffer4] = fbcf(in(n,1),buffer4,Fs,n,...
        d4,g4,amp4,rate4);

    % Combine parallel paths
    combPar = 0.25*(w1 + w2 + w3 + w4);

    % Two series all-pass filters
    [w5,buffer5] = apf(combPar,buffer5,Fs,n,...
        d5,g5,amp5,rate5);

    [out(n,1),buffer6] = apf(w5,buffer6,Fs,n,...
        d6,g6,amp6,rate6);

end

sound(out,Fs);
```

Example 16.8: Script: `schroederReverb.m`

The original algorithm can be further modified to generalize this design approach. An arbitrary number of additional parallel FBCFs with unique delay times can be included to increase the density of the reverb's late reflections. The diffusion characteristics of the algorithm can be changed by including additional APFs in series.

16.3 Moorer

The design approach of the Schroeder algorithm was extended by Moorer (1979). In the Moorer algorithm, two new concepts are added to the parallel FBCFs and series APFs. First, an additional processing block to simulate the early reflections of reverb is included at the input of the algorithm. Second, spectral low-pass filtering is included on the feedback paths of the FBCFs to simulate the roll-off of high frequencies, which occurs in acoustic reverberation.

16.3.1 Early Reflections

The Moorer algorithm includes an additional processing block to simulate early reflections to supplement the FBCFs and APFs (which create the impression of a reverb's late reflections and diffusion, respectively). An example of the block diagram for the Moorer algorithm is shown in Figure 16.6.

The processing block for early reflections is implemented as a **tapped delay line** (TDL). This type of delay effect is created using a single delay buffer. The output of the TDL is created by summing together various *taps* (array elements) at different amplitudes. An example of the TDL used for the early reflections in the Moorer algorithm is shown in Figure 16.7.

If the output of the TDL is created by summing together all of the array elements in the delay buffer, then it is identical to performing convolution with an impulse response of the early reflections. Therefore, the Moorer algorithm can be conceptually considered a hybrid reverb with a convolution component and an algorithmic component.

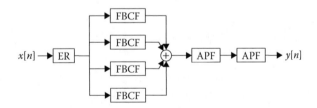

Figure 16.6: Block diagram Moorer reverb

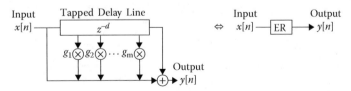

Figure 16.7: Tapped delay line for early reflections

This approach provides the opportunity to create a reverb effect based on a measured impulse response of the early reflections, instead of a hypothetical TDL. In theory, this can make for more realistic results perceptually. Therefore, the Moorer reverb has the benefit of improved sound quality while still maintaining some computational efficiency in the late reflections.

Examples: Create the following function in MATLAB.

```
% EARLYREFLECTIONS
% This function creates a tapped delay line to
% be used for the early reflections of a reverb algorithm.
% The delays and gains of the taps are included in this
% function, and were based on an IR measurement from a
% recording studio in Nashville, TN.
%
% See also MOORERREVERB


function [out,buffer] = earlyReflections(in,buffer,Fs,n)

% Delay times converted from milliseconds
delayTimes = fix(Fs*[0; 0.01277; 0.01283; 0.01293; 0.01333;...
    0.01566; 0.02404; 0.02679; 0.02731; 0.02737; 0.02914; ...
    0.02920; 0.02981; 0.03389; 0.04518; 0.04522; ...
    0.04527; 0.05452; 0.06958]);

% There must be a "gain" for each of the "delayTimes"
gains = [1; 0.1526; -0.4097; 0.2984; 0.1553; 0.1442;...
    -0.3124; -0.4176; -0.9391; 0.6926; -0.5787; 0.5782; ...
    0.4206; 0.3958; 0.3450; -0.5361; 0.417; 0.1948; 0.1548];

% Determine indexes for circular buffer
len = length(buffer);
indexC = mod(n-1,len) + 1; % Current index
buffer(indexC,1) = in;

out = 0; % Initialize the output to be used in loop

% Loop through all the taps
for tap = 1:length(delayTimes)
    % Find the circular buffer index for the current tap
    indexTDL = mod(n-delayTimes(tap,1)-1,len) + 1;
```

```
% "Tap" the delay line and add current tap with output
out = out + gains(tap,1) * buffer(indexTDL,1);


end
```

Example 16.9: Function: `earlyReflections.m`

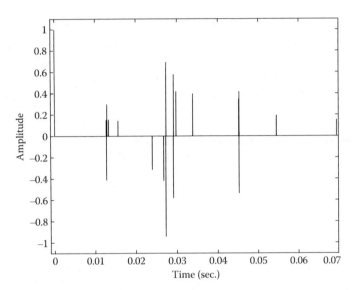

Figure 16.8: Impulse response of an early reflections tapped delay line

The impulse response of an early reflections TDL for the Moorer reverb algorithm is shown in Figure 16.8. The tapped delay line is based on an IR measurement of a recording studio in Nashville, Tennessee. A segment at the beginning of the IR was extracted with sparse reflections. The sample values with the greatest amplitudes were placed in the array of the TDL, along with their corresponding delay times.

16.3.2 Low-Pass Filtered Feedback Delay

Another modification included in the Moorer reverb algorithm is the addition of an LPF in the feedback path of the FBCFs. The purpose of the LPF is to decrease the amplitude of high frequencies as the signal repeats. Conceptually, this is similar to the high-frequency roll-off of sound in air (i.e., a real acoustic space). This additional processing step can give a reverb algorithm more of a natural sound, and less of a metallic sound. An example of the FBCF block diagram is shown in Figure 16.9.

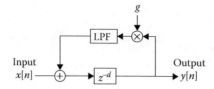

Figure 16.9: Block diagram of low-pass filtered feedback comb filter for Moorer reverb

There are many low-pass filter designs that could be used in the feedback path of the comb filter. Several different designs can be referenced from Chapters 12 and 13. The choice of LPF for implementation can be based on creating a filter with low computational complexity or creating a filter with a similar frequency response to air. A feedback comb filter with a low-pass filter in the feedback path is implemented as a MATLAB function in Example 16.10.

Examples: Create the following function in MATLAB.

```
% LPCF
% This function creates a feedback comb filter
% with an LPF in the feedback path.
%
% Input Variables
%   n : current sample number of the input signal
%   delay : samples of delay
%   fbGain : feedback gain (linear scale)
%   amp : amplitude of LFO modulation
%   rate : frequency of LFO modulation
%   fbLPF : output delayed one sample to create basic LPF
%
% See also MOORERREVERB

function [out,buffer,fbLPF] = lpcf(in,buffer,Fs,n,delay,...
    fbGain,amp,rate,fbLPF)


% Calculate time in seconds for the current sample
t = (n-1)/Fs;
fracDelay = amp * sin(2*pi*rate*t);
intDelay = floor(fracDelay);
frac = fracDelay - intDelay;


% Determine indexes for circular buffer
```

```
len = length(buffer);
indexC = mod(n-1,len) + 1; % Current index
indexD = mod(n-delay-1+intDelay,len) + 1; % Delay index
indexF = mod(n-delay-1+intDelay+1,len) + 1; % Fractional index


out = (1-frac)*buffer(indexD,1) + (frac)*buffer(indexF,1);


% Store the current output in appropriate index
% The LPF is created by adding the current output
% with the previous sample, both are weighted 0.5
buffer(indexC,1) = in + fbGain*(0.5*out + 0.5*fbLPF);


% Store the current output for the feedback LPF
% to be used with the next sample
fbLPF = out;
```

Example 16.10: Function: lpcf.m

16.3.3 *Moorer Algorithm Implementation*

In Example 16.11, a MATLAB implementation of the Moorer reverb algorithm is demonstrated. The block diagram in Figure 16.6 represents the approach to processing.

Examples: Create the following m-file in MATLAB.

```
% MOORERREVERB
% This script implements the Moorer reverb algorithm
% by modifying the Schroeder reverb script. First,
% an additional step to add early reflections is included.
% Second, a simple low-pass filter is included in the feedback
% path of the comb filters.
%
% See also EARLYREFLECTIONS, LPCF


clear;clc;

[in,Fs] = audioread('AcGtr.wav');
in = [in;zeros(Fs*3,1)]; % Add zero-padding for reverb tail
```

```
% Max delay of 70 ms
maxDelay = ceil(.07*Fs);
% Initialize all buffers
buffer1 = zeros(maxDelay,1); buffer2 = zeros(maxDelay,1);
buffer3 = zeros(maxDelay,1); buffer4 = zeros(maxDelay,1);
buffer5 = zeros(maxDelay,1); buffer6 = zeros(maxDelay,1);

% Early reflections tapped delay line
bufferER = zeros(maxDelay,1);

% Delay and gain parameters
d1 = fix(.0297*Fs); g1 = 0.9;
d2 = fix(.0371*Fs); g2 = -0.9;
d3 = fix(.0411*Fs); g3 = 0.9;
d4 = fix(.0437*Fs); g4 = -0.9;
d5 = fix(.005*Fs); g5 = 0.7;
d6 = fix(.0017*Fs); g6 = 0.7;

% LFO parameters
rate1 = 0.6; amp1 = 8;
rate2 = 0.71; amp2 = 8;
rate3 = 0.83; amp3 = 8;
rate4 = 0.95; amp4 = 8;
rate5 = 1.07; amp5 = 8;
rate6 = 1.19; amp6 = 8;

% Variables used as delay for a simple LPF in each comb filter function
fbLPF1 = 0; fbLPF2 = 0; fbLPF3 = 0; fbLPF4 = 0;

% Initialize output signal
N = length(in);
out = zeros(N,1);

for n = 1:N

    % Early reflections TDL
    [w0 , bufferER] = earlyReflections(in(n,1),bufferER,Fs,n);

    % Four parallel LPCFs
    [w1,buffer1,fbLPF1] = lpcf(w0,buffer1,Fs,n,...
```

```
              d1,g1,amp1,rate1,fbLPF1);

    [w2,buffer2,fbLPF2] = lpcf(w0,buffer2,Fs,n,...
        d2,g2,amp2,rate2,fbLPF2);

    [w3,buffer3,fbLPF3] = lpcf(w0,buffer3,Fs,n,...
        d3,g3,amp3,rate3,fbLPF3);

    [w4,buffer4,fbLPF4] = lpcf(w0,buffer4,Fs,n,...
        d4,g4,amp4,rate4,fbLPF4);

    % Combine parallel paths
    combPar = 0.25*(w1 + w2 + w3 + w4);

    % Two series all-pass filters
    [w5,buffer5] = apf(combPar,buffer5,Fs,n,...
        d5,g5,amp5,rate5);

    [out(n,1),buffer6] = apf(w5,buffer6,Fs,n,...
        d6,g6,amp6,rate6);

end

sound(out,Fs);
```

Example 16.11: Script: `moorerReverb.m`

16.4 Feedback Delay Networks

Another algorithm used to simulate reverberation is the **feedback delay network** (FDN). This design further extends the feedback approach introduced by Schroeder (Section 16.2) to create repetitions. In an FDN, each delay block is fed back into itself. Additionally, the output of each delay block can be fed back to the input of any other delay block (Jot, 1992).

Consider conceptually how reflections travel in a room. The front and back walls create one feedback loop, as do the side walls, and the floor and ceiling. However, each feedback loop does not reflect in isolation. A sound wave reflecting between the floor and ceiling may eventually encounter one of the walls and reflect in a new feedback path. The crossover feedback in the FDN simulates this possibility.

16.4.1 Crossover Feedback

In a FDN, there are multiple delay lines in parallel, and each line can have a different delay time, $\{d_1, d_2, \ldots\}$. The output of each delay block is fed back to the input of each delay block. This includes feedback to the same delay line to create a conventional comb filter, as well as crossover feedback between delay lines. Each feedback path may have a separate gain amount as a parameter of the algorithm.

An example of an FDN block diagram with two parallel delay lines is shown in Figure 16.10. The feedback gain from the output of delay block 1 to the input of delay block 1 is labeled g_{11}. In general, the feedback gain from the output of delay block i to the input of delay block j is labeled g_{ij}.

Feedback Matrix

A common way to draw the same diagram shown in Figure 16.10 is to use a matrix to represent the various feedback gains. An equivalent diagram using a matrix is shown in Figure 16.11. The advantage of structuring the diagram with the matrix is to simplify the organization of feedback. This becomes even more important as the number of delay lines increases.

A MATLAB implementation of the block diagram in Figure 16.11 is demonstrated in Example 16.12. Arbitrary delay times and feedback gains were selected. The impulse response of the system is shown in Figure 16.12.

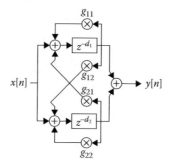

Figure 16.10: Block diagram of parallel comb filters with crossover feedback

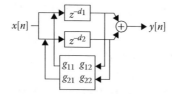

Figure 16.11: Block diagram of parallel comb filters with feedback matrix

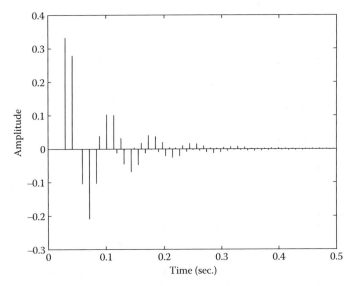

Figure 16.12: Impulse response of parallel comb filters with crossover feedback

Examples: Create the following m-files in MATLAB.

```
% CROSSOVERFEEDBACK
% This script implements two comb filters with
% crossover feedback
%
% See also MODDELAY
clear;clc;


Fs = 48000;
in = [1 ; zeros(Fs*0.5,1)];

% Max delay of 70 ms
maxDelay = ceil(.07*Fs);
% Initialize all buffers
buffer1 = zeros(maxDelay,1); buffer2 = zeros(maxDelay,1);

% Delay and gain parameters
d1 = fix(.0297*Fs);
d2 = fix(.0419*Fs);
g11 = -0.75; g12 = -0.75;
g21 = -0.75; g22 = -0.75;
```

```
% LFO parameters
rate1 = 0.6; amp1 = 3;
rate2 = 0.71; amp2 = 3;

% Initialize output signal
N = length(in);
out = zeros(N,1);


fb1 = 0; fb2 = 0; % feedback holding variables


for n = 1:N

    % Combine input with feedback for respective delay lines
    inDL1 = in(n,1) + fb1;
    inDL2 = in(n,1) + fb2;

    % Two parallel delay lines

    [outDL1,buffer1] = modDelay(inDL1,buffer1,Fs,n,...
    d1,amp1,rate1);

    [outDL2,buffer2] = modDelay(inDL2,buffer2,Fs,n,...
    d2,amp2,rate2);

    % Combine parallel paths
    out(n,1) = 0.5*(outDL1 + outDL2);

    % Calculate feedback (including crossover)
    fb1 = 0.5*(g11 * outDL1 + g21 * outDL2);
    fb2 = 0.5*(g12 * outDL1 + g22 * outDL2);

end
plot(out);
```

Example 16.12: Script: `crossoverFeedback.m`

```
% MODDELAY
% This function creates a series delay effect
% using a buffer. The delay time can be modulated
% based on the LFO parameters "depth" and "rate"
```

```
%
% Input Variables
%   in : single sample of the input signal
%   buffer : used to store delayed samples of the signal
%   n : current sample number used for the LFO
%   depth : range of modulation (samples)
%   rate : speed of modulation (frequency, Hz)


function [out,buffer] = modDelay(in,buffer,Fs,n,...
    delay,depth,rate)


% Calculate time in seconds for the current sample
t = (n-1)/Fs;
fracDelay = depth * sin(2*pi*rate*t);
intDelay = floor(fracDelay);
frac = fracDelay - intDelay;


% Determine indexes for circular buffer
len = length(buffer);
indexC = mod(n-1,len) + 1; % Current index
indexD = mod(n-delay-1+intDelay,len) + 1; % Delay index
indexF = mod(n-delay-1+intDelay+1,len) + 1; % Fractional index



out = (1-frac) * buffer(indexD,1) + ...
    (frac) * buffer(indexF,1);


% Store the current output in appropriate index
buffer(indexC,1) = in;
```

Example 16.13: Function: modDelay.m

16.4.2 *FDN Algorithm Implementation*

The design of an FDN can be generalized to include an arbitrary number of parallel delay paths (Figure 16.13). There are many reverbs with different sounds that can be created by changing the feedback gains, $\{g_{11}, g_{12}, \ldots\}$. A version of the Schroeder algorithm can be created using an FDN by setting all feedback gains in the matrix to zero, except for $g_{11}, g_{22}, g_{33}, g_{44}$, etc.

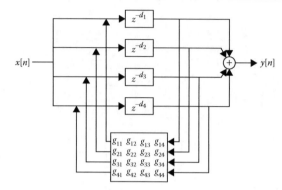

Figure 16.13: Block diagram of feedback delay network with four delay lines

In addition to the feedback gains, each delay path could be scaled in amplitude to change their relative level. For an implementation of the FDN with a variable parameter to control reverb time, a single value can be used to scale the amplitude of all feedback paths. An implementation of an FDN reverb effect is shown in Example 16.14.

Examples: Create the following m-file in MATLAB.

```
% FDNEXAMPLE
% This script implements a feedback delay network
% using a Stautner and Puckette matrix.
%
% See also MODDELAY
clear;clc;

[in,Fs] = audioread('AcGtr.wav');
% Add extra space at end for the reverb tail
in = [in;zeros(Fs*3,1)];

% Max delay of 70 ms
maxDelay = ceil(.07*Fs);
% Initialize all buffers
buffer1 = zeros(maxDelay,1); buffer2 = zeros(maxDelay,1);
buffer3 = zeros(maxDelay,1); buffer4 = zeros(maxDelay,1);

% Delay and gain parameters
d1 = fix(.0297*Fs);
d2 = fix(.0371*Fs);
d3 = fix(.0411*Fs);
```

```
d4 = fix(.0437*Fs);
g11 = 0; g12 = 1; g13 = 1; g14 = 0;   % Stautner and Puckette
g21 =-1; g22 = 0; g23 = 0; g24 =-1;   % feedback matrix
g31 = 1; g32 = 0; g33 = 0; g34 =-1;
g41 = 0; g42 = 1; g43 =-1; g44 = 0;


% LFO parameters
rate1 = 0.6; amp1 = 5;
rate2 = 0.71; amp2 = 5;
rate3 = 0.83; amp3 = 5;
rate4 = 0.95; amp4 = 5;


% Initialize output signal
N = length(in);
out = zeros(N,1);


% Feedback holding variables
fb1 = 0; fb2 = 0; fb3 = 0; fb4 = 0;


% Gain to control reverb time
g = .67;
for n = 1:N

    % Combine input with feedback for respective delay lines
    inDL1 = in(n,1) + fb1;
    inDL2 = in(n,1) + fb2;
    inDL3 = in(n,1) + fb3;
    inDL4 = in(n,1) + fb4;

    % Four parallel delay lines
    [outDL1,buffer1] = modDelay(inDL1,buffer1,Fs,n,...
    d1,amp1,rate1);

    [outDL2,buffer2] = modDelay(inDL2,buffer2,Fs,n,...
    d2,amp2,rate2);

    [outDL3,buffer3] = modDelay(inDL3,buffer3,Fs,n,...
    d3,amp3,rate3);

    [outDL4,buffer4] = modDelay(inDL4,buffer4,Fs,n,...
```

```
     d4,amp4,rate4);

     % Combine parallel paths
     out(n,1) = 0.25*(outDL1 + outDL2 + outDL3 + outDL4);

     % Calculate feedback (including crossover)
     fb1 = g*(g11*outDL1 + g21*outDL2 + g31*outDL3 + g41*outDL4);
     fb2 = g*(g12*outDL1 + g22*outDL2 + g32*outDL3 + g42*outDL4);
     fb3 = g*(g13*outDL1 + g23*outDL2 + g33*outDL3 + g43*outDL4);
     fb4 = g*(g14*outDL1 + g24*outDL2 + g34*outDL3 + g44*outDL4);

end
sound(out,Fs);
```

Example 16.14: Script: `fdnExample.m`

In Example 16.14, the feedback matrix is based on the Stautner and Puckette (1982) design:

$$\begin{bmatrix} 0 & 1 & 1 & 0 \\ -1 & 0 & 0 & -1 \\ 1 & 0 & 0 & -1 \\ 0 & 1 & -1 & 0 \end{bmatrix}$$

Other common examples of feedback matrices include the Hadamard and Householder designs (Smith, 2010). The Hadamard design uses the following matrix:

$$\begin{bmatrix} 1 & 1 & 1 & 1 \\ 1 & -1 & 1 & -1 \\ 1 & 1 & -1 & -1 \\ 1 & -1 & -1 & 1 \end{bmatrix}$$

The Householder design uses the following matrix:

$$\begin{bmatrix} 1 & -1 & -1 & -1 \\ -1 & 1 & -1 & -1 \\ -1 & -1 & 1 & -1 \\ -1 & -1 & -1 & 1 \end{bmatrix}$$

16.5 Appendix: Estimating Reverb Time

The length of reverb is measured as the time it takes for a sound reflecting in a room to decrease by 60 decibels. Reverb time is called RT-60. A basic way to estimate RT-60 is based on visualizing an impulse response on the decibel scale (Figure 16.14).

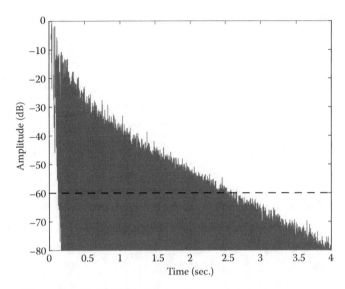

Figure 16.14: RT-60 estimate of a reverb impulse response

Examples: Create the following m-file in MATLAB.

```
% RT60
% This script analyzes the RT-60
% of a feedback delay network by visualizing
% an impulse response on the decibel scale.
clear;clc;

Fs = 48000; Ts = 1/Fs;
in = [ 1 ; zeros(5*Fs,1)]; % Impulse signal

% Max delay of 70 ms
maxDelay = ceil(.07*Fs);
% Initialize all buffers
buffer1 = zeros(maxDelay,1); buffer2 = zeros(maxDelay,1);
buffer3 = zeros(maxDelay,1); buffer4 = zeros(maxDelay,1);
```

```
% Delay and gain parameters
d1 = fix(.0297*Fs);
d2 = fix(.0371*Fs);
d3 = fix(.0411*Fs);
d4 = fix(.0437*Fs);
g11 = 0; g12 = 1; g13 = 1; g14 = 0;
g21 =-1; g22 = 0; g23 = 0; g24 =-1;
g31 = 1; g32 = 0; g33 = 0; g34 =-1;
g41 = 0; g42 = 1; g43 =-1; g44 = 0;

% LFO parameters
rate1 = 0.6; amp1 = 5;
rate2 = 0.71; amp2 = 5;
rate3 = 0.83; amp3 = 5;
rate4 = 0.95; amp4 = 5;

% Initialize output signal
N = length(in);
out = zeros(N,1);

% Feedback holding variables
fb1 = 0; fb2 = 0; fb3 = 0; fb4 = 0;

% Gain to control reverb time
g = .67;
for n = 1:N

    % Combine input with feedback for respective delay lines
    inDL1 = in(n,1) + fb1;
    inDL2 = in(n,1) + fb2;
    inDL3 = in(n,1) + fb3;
    inDL4 = in(n,1) + fb4;

    % Four parallel delay lines
    [outDL1,buffer1] = modDelay(inDL1,buffer1,Fs,n,...
    d1,amp1,rate1);

    [outDL2,buffer2] = modDelay(inDL2,buffer2,Fs,n,...
    d2,amp2,rate2);
```

```
    [outDL3,buffer3] = modDelay(inDL3,buffer3,Fs,n,...
    d3,amp3,rate3);


    [outDL4,buffer4] = modDelay(inDL4,buffer4,Fs,n,...
    d4,amp4,rate4);


    % Combine parallel paths
    out(n,1) = 0.25*(outDL1 + outDL2 + outDL3 + outDL4);


    % Calculate feedback (including crossover)
    fb1 = g*(g11*outDL1 + g21*outDL2 + g31*outDL3 + g41*outDL4);
    fb2 = g*(g12*outDL1 + g22*outDL2 + g32*outDL3 + g42*outDl4);
    fb3 = g*(g13*outDL1 + g23*outDL2 + g33*outDL3 + g43*outDL4);
    fb4 = g*(g14*outDL1 + g24*outDL2 + g34*outDL3 + g44*outDL4);


end
out = out/max(abs(out)); % Normalize to unity gain (0 dB)


t = [0:N-1]*Ts;
plot(t,20*log10(abs(out)));
line([0 4],[-60 -60],'Color','red','LineStyle','--');
axis([0 4 -80 0]);
```

Example 16.15: Script: rt60.m

Bibliography

Jot, J.-M. (1992). *Etude et réalisation d'un spatialisateur de sons par modèles physiques et perceptifs*. PhD thesis, Télécom Paris, France.

Moorer, J. A. (1979). About this reverberation business. *Computer Music Journal, 3*(2), 13–18.

Schroeder, M. R. (1962). Natural-sounding artificial reverberation. *Journal of the Audio Engineering Society, 10*(3), 219–223.

Smith, J. O. (2010). *Physical audio signal processing: for virtual musical instruments and digital audio effects*, W3K Publishing. Retrieved from www.books.w3k.org, ISBN 978-0-9745607-2-4.

Stautner, J., & Puckette, M. (1982). Designing multi-channel reverberators. *Computer Music Journal, 6*(1), 52–65.

Chapter 17

Amplitude Envelope Effects
Envelope Follower, Vocoder, and Transient Designer

17.1 Introduction: Amplitude Envelope Effects

This chapter introduces several audio effects based on a signal's amplitude envelope. These effects combine many other processors including: spectral processors (Chapters 12 and 13), nonlinear processors (Chapter 10), summing, and amplitude modulation (Chapter 8). Additionally, this chapter provides the foundation for the processing used in the dynamic range effects described in Chapter 18.

17.1.1 Defining the Amplitude Envelope

A signal's **amplitude envelope** is the overall gross shape of how amplitude changes over time. In some cases, the amplitude envelope is referred to as the temporal envelope to specifically distinguish it from a spectral envelope or curve.

The amplitude envelope describes the slow-varying fluctuations in amplitude, and is meant to be independent of the relatively high-frequency pitch. As an example, the amplitude envelope for different notes played on a piano is similar, regardless of the fundamental frequency. The amplitude envelope for a staccato note has a fast attack time and a fast decay time. Conversely, a bowed legato note on a violin has a slow attack time and long sustain time. In general, an amplitude envelope is an approximation to the attack, decay, sustain, and release (ADSR) for an instrument.

17.1.2 Effects Based on the Amplitude Envelope

There are several different audio effects based on the amplitude envelope. The ADSR of a synthesizer is controlled by shaping the amplitude envelope. All types of envelope-follower effects are based on extracting the amplitude envelope from one signal and applying it to something else—ranging from another signal's amplitude, to a filter's cutoff frequency (envelope wah), to a reverb's time. One special type of envelope-follower effect is the vocoder, in which the amplitude envelope from separate frequency bands is used to modulate the amplitude of another signal. The final effect described in this chapter is the transient designer, which can be used to increase or decrease the attack and sustain of a signal.

17.2 Signal Synthesis: ADSR

An important part of musical signal synthesis is shaping the ADSR amplitude envelope of the carrier signal. In doing so, basic test signals (Chapter 7) can be transformed into sounds ranging from synthesized snare drums, to plucked guitar strings, to legato flute notes. This process requires two steps. First, the ADSR amplitude envelope needs to be created. Second, it is applied to the carrier signal.

17.2.1 Creating the ADSR Amplitude Envelope

The ADSR amplitude envelope consists of four sections. One approach to creating the enveleope is to concatenate four separate amplitude curves together. Each individual curve can be created by considering it as an amplitude fade (Section 8.4). All types of fades can be used including linear, exponential, etc. An example of an ADSR amplitude envelope based on linear fades is shown in Figure 17.1.

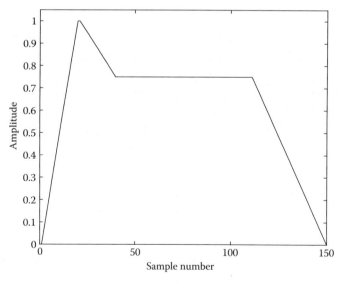

Figure 17.1: ADSR amplitude envelope

Examples: Create the following m-file in MATLAB.

```
% ADSREXAMPLE
% This script creates a linear ADSR amplitude envelope
%
% See also ADSR

% Number of samples per fade
a = 20;
d = 20;
s = 70;
r = 40;

sustainAmplitude = 0.75;

% Create each segment A,D,S,R
aFade = linspace(0,1,a)';
dFade = linspace(1,sustainAmplitude,d)';
sFade = sustainAmplitude * ones(s,1);
rFade = linspace(sustainAmplitude,0,r)';


% Concatenates total ADSR envelope
env = [aFade;dFade;sFade;rFade];

plot(env);
```

Example 17.1: Script: `adsrExample.m`

17.2.2 Applying the ADSR Amplitude Envelope

The process of applying the ADSR amplitude envelope to the carrier signal during synthesis is based on the element-wise multiplication operation. The amplitude of each sample of the carrier signal, $x[n]$, is scaled by the amplitude of the same time sample in the ADSR curve, $g_{env}[n]$. A block diagram of the process is shown in Figure 17.2.

In Example 17.2, a MATLAB function is shown that processes an input signal by applying an ADSR amplitude envelope based on parameters including attack time, decay time, release time, and sustain amplitude. The envelope is created to be the same number of samples as the input signal for element-wise multiplication.

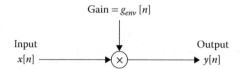

Figure 17.2: Element-wise multiplication to apply an ADSR amplitude envelope

Examples: Create the following m-file in MATLAB.

```
% ADSR
% This function can be used to apply an
% ADSR envelope on to an input signal.
%
% Input Variables
%    attackTime : length of attack ramp in milliseconds
%    decayTime : length of decay ramp in ms
%    sustainAmplitude : linear amplitude of sustain segment
%    releaseTime : length of release ramp in ms

function [ y ] = adsr( x,Fs,attackTime,decayTime,...
    sustainAmplitude,releaseTime)

% Convert time inputs to seconds
attackTimeS = attackTime / 1000;
decayTimeS = decayTime / 1000;
releaseTimeS = releaseTime / 1000;

% Convert seconds to samples and determine sustain time
a = round(attackTimeS * Fs);      % Round each to an integer
d = round(decayTimeS * Fs);       % number of samples
r = round(releaseTimeS * Fs);
s = length(x) - (a+d+r);          % Determine length of sustain

% Create linearly spaced fades for A. D. and R. Create hold for S
aFade = linspace(0,1,a)';
dFade = linspace(1,sustainAmplitude,d)';
sFade = sustainAmplitude * ones(s,1);
rFade = linspace(sustainAmplitude,0,r)';

% Concatenates total ADSR curve
```

```
adsr = [aFade;dFade;sFade;rFade];


% Applies ADSR shaping to X
y = x .* adsr;


end
```

Example 17.2: Function: `adsr.m`

17.3 Measuring the Amplitude Envelope

For many audio effects, the amplitude envelope is not synthesized. Rather, it is measured from the waveform of a signal itself. Therefore, the amplitude envelope is based on the recording of an actual snare drum, plucked guitar string, sustained flute, etc. The aim is to measure an amplitude envelope similar in shape to the ADSR curve, which can then be applied to another signal using an element-wise operation.

There are several methods to measure the amplitude envelope from a signal's waveform. In all cases, the envelope is strictly positive (i.e., never has a negative value). This is called a unipolar signal, as opposed to a bipolar signal, which can oscillate between positive and negative numbers. Each method provides an approximate estimate of the overall shape of the waveform.

17.3.1 Linear Method

The linear method of measuring an amplitude envelope is based on processing the waveform directly, without converting it to a different scale first. In this method, the waveform is processed in two sequential stages. First, full-wave rectification (absolute value function, Section 10.4.2) is performed to convert the bipolar waveform to a unipolar signal. Second, low-pass filtering is performed to remove high-frequency oscillations from the signal. The result is an amplitude envelope, $\tilde{x}[n]$, approximating the overall shape of the waveform, $x[n]$. If necessary, make-up gain can be applied to the output signal to scale the amplitude.

A block diagram for the processing is shown in Figure 17.3. A plot of the stages of processing for an acoustic guitar recording is shown in Figure 17.4.

Amplitude Envelope Low-Pass Filtering

The LPF used as part of the process to measure an amplitude envelope is included to smooth over the overall shape. When used with musical signals, the LPF is designed to attenuate the

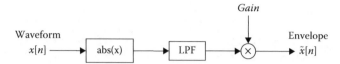

Figure 17.3: Block diagram of linear method to measure an amplitude envelope

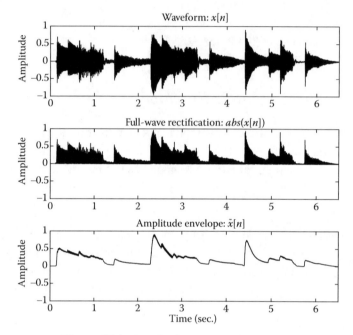

Figure 17.4: Amplitude envelope processing

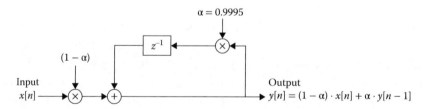

Figure 17.5: Block diagram of LPF used to measure an amplitude envelope

fundamental frequency and higher harmonics in the signal. Therefore, the cutoff frequency of the filter is chosen to be less than ∼100 Hz.

Filter design strategies (Chapters 12 and 13) can be used to design LPFs with various orders and complexity. A basic IIR system for low-pass filtering the amplitude envelope is shown in Figure 17.5. The frequency response of the system is shown in Figure 17.6. The filter attenuates all frequencies, except those close to ∼0 Hz.

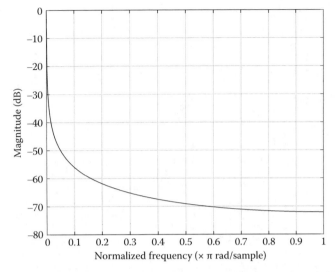

Figure 17.6: Magnitude response of the amplitude envelope LPF

Examples: Execute the following commands in MATLAB.

```
% This script plots the frequency response
% of the LPF used in an amplitude envelope analysis
alpha = 0.9995;    % Gain on the feedback path
b = [1-alpha];     % Feedforward coefficients
a = [1 , -alpha];  % Feedback coefficients
freqz(b,a);        % Frequency response of the LPF
```

Example 17.3: Script: `feedbackLPF.m`

The amount of smoothing performed by the LPF on the amplitude envelope is set by the gain parameter, α. To achieve adequate smoothing, α is selected within a range of [0.999, 0.9999]. As α increases, the cutoff frequency of the LPF decreases, and greater smoothing is performed. A comparison of two values of α is shown in Figure 17.7 for the amplitude envelope measured from the waveform in Figure 17.4.

17.3.2 Decibel Method

For some audio effects that use a signal's amplitude envelope, the measurement is performed relative to the decibel scale. Therefore, it is necessary to convert a signal's amplitude from the linear scale to the decibel scale as part of the process. After the envelope smoothing is complete, the decibel amplitude is converted back to the linear scale. A block diagram of the amplitude envelope measurement is shown in Figure 17.8.

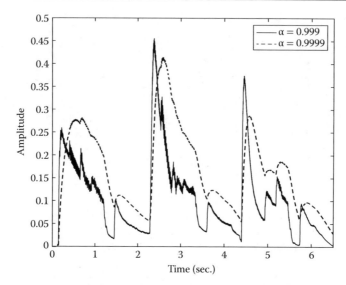

Figure 17.7: Amplitude envelope LPF comparison

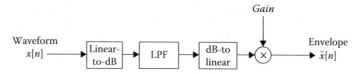

Figure 17.8: Block diagram of decibel method to measure an amplitude envelope

To convert an amplitude value from the linear scale to the decibel scale, the following expression is used: $20 \cdot log_{10}(|x[n]|)$. The result converts the original bipolar signal (ranging from −1 to 1 FS) to a unipolar signal (ranging from −∞ to 0 dBFS). An example of performing this conversion for a sine-wave signal is shown in Figure 17.9.

In Figure 17.10, a comparison is shown of the linear and decibel approaches to measuring a signal's amplitude envelope.

17.3.3 RMS Method

A third approach to measuring an amplitude envelope uses an RMS calculation (Section 6.7.3). Here, the measurement is based on a conceptual *average* amplitude rather than the smoothed instantaneous, or *peak*, amplitude.

Calculating the RMS value for the entire signal would provide a single scalar representing the overall strength. For an amplitude envelope, a measurement of the signal strength is necessary

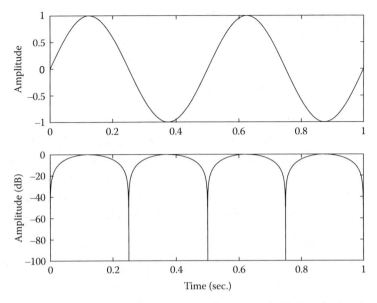

Figure 17.9: Converting a linear, bipolar signal to a decibel, unipolar signal

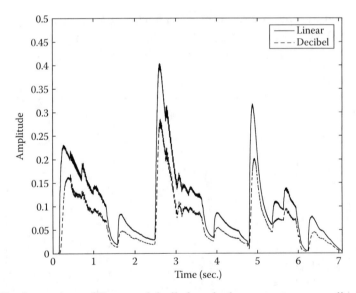

Figure 17.10: Comparison of linear and decibel methods to measure an amplitude envelope

for every sample. Therefore, the RMS calculation is not performed on the entire signal. Rather, the RMS calculation is performed within a short time frame surrounding each sample. This creates an estimate of the average signal strength close to a given sample. The following expression can be used to calculate a signal's RMS amplitude at sample number n (i.e., $x_{rms}[n]$) within a time frame of length M samples:

$$(x_{rms}[n])^2 = \frac{1}{M} \sum_{m=-\frac{M}{2}}^{\frac{M}{2}-1} (x[n+m])^2$$

This RMS calculation creates a strictly positive envelope, and also inherently performs smoothing through the averaging process. Therefore, it is unnecessary to low-pass filter the measured values in this approach (although it could be included as an additional step). The number of included samples, M, for the RMS calculation can be set to change the length of the time frame for the calculation. A longer time frame results in a smoother envelope.

In Figure 17.11, a comparison is shown of the measured amplitude envelope for the linear and RMS methods. In Figure 17.12, two measured amplitude envelopes are shown with different time frame sizes, M, for the RMS method.

17.4 Envelope Amplitude Modulation

As was demonstrated with the ADSR envelope in Section 17.2.2, one way an amplitude envelope can be applied to another signal is by modulating the amplitude. This is performed as an element-wise multiplication of the amplitude envelope and the signal.

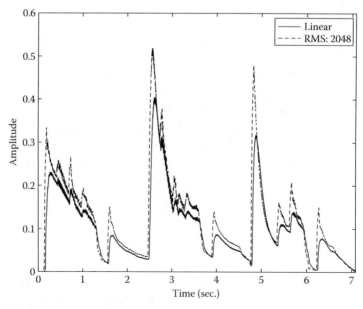

Figure 17.11: Comparison of linear and RMS methods to measure an amplitude envelope

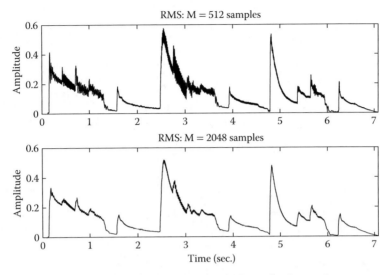

Figure 17.12: Changing the frame size for the RMS amplitude envelope measurement

In Example 17.4, the amplitude envelope is measured from the waveform of a voice recording. Then the envelope is used to modulate the amplitude of an electronic synthesizer recording. The sound files demonstrated in the example are included in the book's *Additional Content*.

Examples: Create and run the following m-file in MATLAB.

```
% ENVELOPEMODULATION
% This script demonstrates the process of measuring an
% amplitude envelope from the waveform of a voice recording
% and using it to modulate the amplitude of synth recording.

% Import audio files
[in,Fs] = audioread('Voice.wav');
[synth] = audioread('Synth.wav');

alpha = 0.9997;   % Feedback gain
fb = 0;           % Initialized value for feedback
N = length(in);
env = zeros(N,1);
for n = 1:N
    % Analyze envelope
    env(n,1) = (1-alpha) * abs(in(n,1)) + alpha * fb;
```

```
        fb = env(n,1);

end
% Make-up gain
env = 4*env;

% Amplitude modulation of envelope applied to synthsizer
out = synth.*env;

sound(out,Fs);
```

Example 17.4: Script: envelopeModulation.m

17.5 Vocoder

The **vocoder** is an envelope-follower effect that uses amplitude modulation. This effect extends the process of measuring and applying an amplitude envelope to multiple frequency regions of a signal.

The vocoder effect has been used in electronic music to create a robotic voice. Conventionally, a signal of a human voice is combined with a signal of an electronic synthesizer. The process of combining the two signals involves first filtering each signal into multiple frequency bands. The filtering is performed by a filter bank to create the multi-band processing (Section 13.5.2).

17.5.1 Multi-Band Envelope Modulation

To create the vocoder effect, amplitude modulation is performed using the envelope measured in each frequency region. Both the voice signal and synthesizer signal are routed separately through identical filterbanks. Then, the envelope is measured on each band of the voice signal and used to modulate the amplitude of each band of the synthesizer signal. Finally, all the frequency bands are summed together to create the output vocoded signal. A block diagram of the processing is shown in Figure 17.13.

Examples: Create and run the following m-file in MATLAB.

```
% VOCODEREXAMPLE
% This script demonstrates the process of creating
% a vocoder effect using a voice signal and
% synth signal
```

```matlab
clc;clear;close all;


% Import audio files
[in,Fs] = audioread('Voice.wav');
[synth] = audioread('Synth.wav');

% Initialize filter parameters
Nyq = Fs/2;    % Nyquist frequency
order = 2;     % Filter order

numOfBands = 16;

% Logarithmically space cutoff frequencies
% 2*10^1 - 2*10^4 (20-20k) Hz
freq = 2 * logspace(1,4,numOfBands+1);

g = 0.9992; % Smoothing filter gain
fb = 0;     % Initialize feedback delay

N = length(in);

% These arrrays are used to store the filtered
% versions of the input signal. Each column stores
% the signal for each band. As an example,
% voxBands(:,4) stores the band-pass filtered
% signal in the fourth band.
voxBands = zeros(N,numOfBands);
synthBands = zeros(N,numOfBands);
envBands = zeros(N,numOfBands);

for band = 1:numOfBands % Perform processing 1 band per loop

    % Determine lower and upper cutoff frequencies
    % of the current BPF on a normalized scale.
    Wn = [freq(band) , freq(band+1)] ./ Nyq;
    [b,a] = butter(order,Wn);

    % Filter signals and store the result
    voxBands(:,band) = filter(b,a,in);
```

```
        synthBands(:,band) = filter(b,a,synth);

        % Envelope measurement from vocal signal
        for n = 1:N

            envBands(n,band) = (1-g) * abs(voxBands(n,band)) + g * fb;
            fb = envBands(n,band);

        end
        fb = 0;

    end

    % Perform amplitude modulation
    outBands = zeros(length(in),numOfBands);
    for band = 1:numOfBands

        % Apply the envelope of the vocal signal to the synthesizer
        % in each of the bands
        outBands(:,band) = envBands(:,band) .* synthBands(:,band);

    end

    % Sum together all the bands
    out = sum(outBands,2);
    % Make-up gain
    out = 32 * out;

    % Listen to the output and plot it
    sound(out,Fs); plot(out);
```

Example 17.5: Script: `vocoderExample.m`

17.5.2 Cochlear Implant Simulation

One use of the vocoding process is to simulate cochlear implants. Vocoded speech, also called **chimeric speech**, was proposed for this application in Shannon, Zeng, Kamath, Wygonski, and Ekelid (1995). As a substitute for an electronic synthesizer, white noise is used as the input carrier signal.

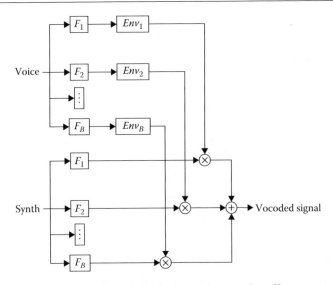

Figure 17.13: Block diagram of a vocoder effect

Background

Cochlear implants are audio devices given to listeners with profound hearing loss who would not benefit from the acoustic amplification of hearing aids. The device is surgically implanted on a recipient, including the insertion of an electrode array into their cochlea. The electrode array stimulates the nerves connecting to the cochlea in ~10–16 regions (or locations). This is important because different locations in the cochlea primarily encode different frequencies.

The basic processing of the device is the following. First, a microphone captures and converts acoustic sound to electricity, which is then digitally sampled. The digital signal is processed by a filterbank to separate each frequency band. An envelope of each frequency band is measured and used to electrically stimulate different electrodes on the array.

To investigate how cochlear implants should function, audio engineers simulate a model of how it might sound for listeners with normal hearing. To re-create the degraded spectral resolution of ~10–16 bands, a filterbank is used to separate white noise into different regions. Then, the amplitude envelope of speech is used to modulate each band of filtered noise. An implementation of the cochlear implant simulation using a vocoder is demonstrated in Example 17.6.

Examples: Create and run the following m-file in MATLAB.

```
% CISIMULATION
% This script performs vocoding
```

```matlab
% using speech and white noise. This process
% is used to simulate cochlear implants
% for listeners with acoustic hearing
%
% See also VOCODEREXAMPLE

clc;clear;close all;

% Import audio files
[in,Fs] = audioread('Voice.wav');
N = length(in);
noise = 0.1 * randn(N,1);

% Initialize filter parameters
Nyq = Fs/2;
order = 2; % Filter order

numOfBands = 16;

% Logarithmically space cutoff frequencies
% 2*10^1 - 2*10^4 (20-20k) Hz
freq = 2 * logspace(1,4,numOfBands+1);

g = 0.9992; % Smoothing filter gain
fb = 0;     % Initialize feedback delay

voxBands = zeros(N,numOfBands);
noiseBands = zeros(N,numOfBands);
envBands = zeros(N,numOfBands);

for band = 1:numOfBands

    Wn = [freq(band) , freq(band+1)] ./ Nyq;
    [b,a] = butter(order,Wn);

    % Filterbank
    voxBands(:,band) = filter(b,a,in);
    noiseBands(:,band) = filter(b,a,noise);

    % Envelope measurement
```

```
        for n = 1:N

            envBands(n,band) = (1-g) * abs(voxBands(n,band)) + g * fb;
            fb = envBands(n,band);

        end
        fb = 0;

    end

    % Perform amplitude modulation
    outBands = zeros(length(in),numOfBands);
    for band = 1:numOfBands

        outBands(:,band) = envBands(:,band) .* noiseBands(:,band);

    end

    % Sum together all the bands
    out = sum(outBands,2);
    % Make-up gain
    out = 32 * out;

    sound(out,Fs); plot(out);
```

Example 17.6: Script: `ciSimulation.m`

17.6 *Envelope Parameter Modulation*

Besides using the envelope to modulate the amplitude of another signal, it can be used to modulate the parameters of audio effects. One example would be to modulate the cutoff frequency of an LPF based on the envelope.

17.6.1 *Envelope-Wah Effect*

As discussed in Section 15.8.1, the wah-wah effect is based on a resonant LPF with a changing cutoff frequency. The auto-wah effect uses an LFO to modulate the cutoff frequency based on the repetitions of a periodic signal. The **envelope-wah** effect uses the amplitude envelope of a signal to modulate the cutoff frequency.

Examples: Create and run the following m-file in MATLAB.

```matlab
% ENVWAHEXAMPLE
% This script implements an env-wah effect using
% a bi-quad filter as the resonant LPF after
% analyzing the amplitude envelope of the input
% signal.
%
% See also BIQUADWAH

clear;clc;
[x,Fs] = audioread('AcGtr.wav');
Ts = 1/Fs;
N = length(x);
% Initialize feedforward delay buffer
% (stores 2 previous samples of input)
ff = [0 ; 0]; % ff(n,1) = n samples of delay

% Initialize feedback delay buffer
% (stores 2 previous samples of output)
fb = [0 ; 0]; % fb(n,1) = n samples of delay

% Bandwidth of resonant LPF
Q = 4;

% Wet/dry mix
wet = 100;

% Initialize output signal
y = zeros(N,1);
% Cutoff frequency from envelope
cutoff = zeros(N,1);

% Envelope LPF parameters
alpha = 0.9995;
envPreviousValue = 0;

for n = 1:N
    % Envelope detection
```

```
        rect = abs(x(n,1));
        env = (1-alpha) * rect + (alpha)*envPreviousValue;
        envPreviousValue = env;

        % Scale envelope for cutoff frequency of LPF
        freq = 1500 + 10000 * env;

        % Use bi-quad wah effect function
        [y(n,1),ff,fb] = biquadWah(x(n,1),Fs,...
            freq,Q,ff,fb,wet);
        % Store for plotting
        cutoff(n,1) = freq;
    end

sound(y,Fs);

t= [0:N-1]*Ts;

subplot(2,1,1);
plot(t,y);
xlabel('Time (sec.)');ylabel('Amplitude');
subplot(2,1,2);
plot(t,cutoff);
xlabel('Time (sec.)');ylabel('Cutoff Freq. (Hz)');
```

Example 17.7: Script: `envWahExample.m`

By scaling the values of the envelope to a range of frequencies in Hz, it can be used to modulate the cutoff of the LPF. An example is shown in Figure 17.14 of a filter's cutoff frequency based on the amplitude envelope.

17.7 Transient Designer

Another type of envelope effect is the **transient designer**, also called the transient shaper. This effect makes it possible to change the attack and sustain characteristics of a signal's envelope. The effect performs an analysis of the input signal, and uses the analysis to process the input signal. The transient designer is conventionally used with percussive instruments to enhance the signal's transient.

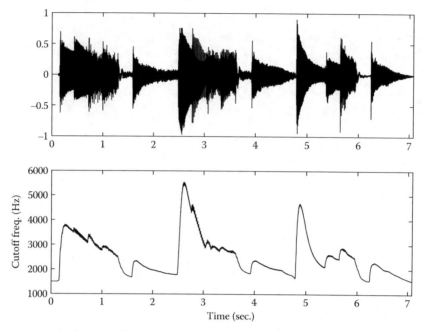

Figure 17.14: Scaling the amplitude envelope to a frequency range in Hz

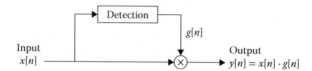

Figure 17.15: Block diagram of a transient designer effect

17.7.1 Envelope Detection Analysis

An important part in creating the transient designer effect is the analysis of the input signal's envelope. The purpose of the analysis is to detect whether the signal's envelope is in an attack stage or a sustain stage. The result of the detection process is a gain scalar, $g[n]$, to increase or decrease the amplitude of the input signal. A block diagram with the parallel detection path is shown in Figure 17.15.

The **detection block** on the parallel path of the effect consists of a comparison between two amplitude envelopes. One envelope is measured with a faster response to closely follow the transients of the signal. The second envelope is measured with a slower, smoother response, such that the transients of the signal are less apparent. Then, the smoother envelope is subtracted from the other, and the **difference envelope** is created (Figure 17.16).

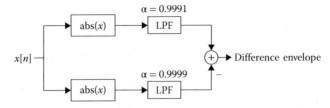

Figure 17.16: Block diagram to produce the transient designer difference envelope

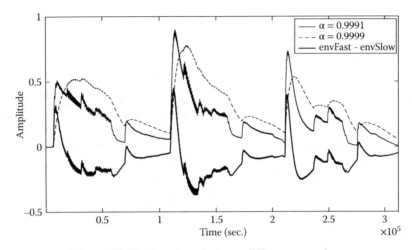

Figure 17.17: Transient designer difference envelope

The analysis on the detection path of the transient designer is demonstrated in Example 17.8. The comparison used to create the difference envelope is shown in Figure 17.17.

Examples: Create and run the following m-file in MATLAB.

```
% TRANSIENTANALYSIS
% This script plots the amplitude envelopes used
% in the transient designer effect. A comparison
% is plotted of the "fast" and "slow" envelopes
% used to determine when "attack" and "sustain"
% is occurring in the signal.
%
% See also TRANSIENTDESIGNER, TRANSIENTEXAMPLE
clear; clc;

[in,Fs]=audioread('AcGtr.wav');

gFast = 0.9991; % Gain smoothing for the "fast" envelope
```

```
fbFast = 0;    % Feedback for the "fast" envelope
gSlow = 0.9999; % Gain smoothing for the "slow" envelope
fbSlow = 0;    % Feedback for the "slow" envelope
N = length(in);
envFast = zeros(N,1);
envSlow = zeros(N,1);
transientShaper = zeros(N,1);
for n = 1:N

    envFast(n,1) = (1-gFast) * 2 * abs(in(n,1)) + gFast * fbFast;
    fbFast = envFast(n,1);

    envSlow(n,1) = (1-gSlow) * 3 * abs(in(n,1)) + gSlow * fbSlow;
    fbSlow = envSlow(n,1);

    transientShaper(n,1) = envFast(n,1) - envSlow(n,1);

end

figure(1);
plot(envFast); hold on;
plot(envSlow);
plot(transientShaper);
hold off;
legend({'$\alpha$ = 0.9991','$\alpha$ = 0.9999', ...
    'envFast - envSlow'},'Interpreter','latex','FontSize',14);
axis([1 length(in) -0.5 1]);

attack = zeros(N,1);
sustain = zeros(N,1);
for n = 1:N

    if transientShaper(n,1) > 0

        attack(n,1) = transientShaper(n,1) + 1;
        sustain(n,1) = 1;
    else

        attack(n,1) = 1;
        sustain(n,1) = transientShaper(n,1) + 1;
```

```
    end

end

figure(2);
subplot(2,1,1);  % Plot the detected attack envelope
plot(attack); title('Attack Envelope', 'FontSize',14);
axis([1 length(in) 0.5 1.5]);
subplot(2,1,2);  % Plot the detected sustain envelope
plot(sustain); title('Sustain Envelope', 'FontSize',14);
axis([1 length(in) 0.5 1.5]);
```

Example 17.8: Script: `transientAnalysis.m`

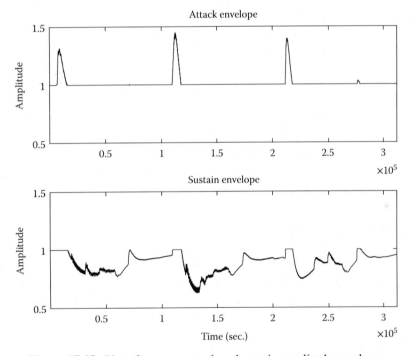

Figure 17.18: Plot of separate attack and sustain amplitude envelopes

When the difference envelope has a positive amplitude, the faster envelope has a greater amplitude than the slower envelope. This result corresponds to an instance of *attack*. When the difference envelope has a negative amplitude, the slower envelope has a greater amplitude than the faster envelope. This result corresponds to an instance of *sustain*. These two mutually exclusive conditions can be separated into independent envelopes (Figure 17.18)

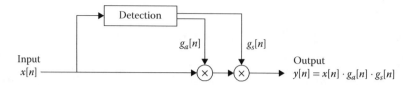

Figure 17.19: Block diagram of separate attack and sustain amplitude envelopes

to be applied to the original, input signal. Finally, an offset of 1 is added to the envelopes. This causes the modulation envelope, $g[n]$, to be unity gain (instead of zero) when the effect is not changing the input signal.

By separating the attack and sustain portions of the amplitude envelope, an audio engineer can have independent control of their effect on a signal. A block diagram of this approach is shown in Figure 17.19

17.7.2 Controls

There are two parameters on the transient designer effect. An *attack* parameter can be used to either increase or decrease the amplitude of a signal's transient. A separate *sustain* parameter can be used to either increase or decrease the amplitude of a signal's sustain.

Both parameters span a normalized range from -1 to 1. When the parameters have a value of 0, the input signal passes through the effect unprocessed. When a parameter has a positive value, the amplitude of the input signal is increased during the appropriate region of the signal. Conversely, when a parameter has a negative value, the amplitude of the input signal is decreased during the appropriate region of the signal.

17.7.3 Transient Designer Implementation

In Example 17.9, an implementation of the transient designer effect is demonstrated. Separate parameters for attack and sustain are passed into a function as variable arguments to adjust the performance of the effect. Several conditions of increased or decreased amplitude are shown in Figures 17.20 through 17.23.

Examples: Create the following m-files in MATLAB.

```
% TRANSIENTEXAMPLE
% This script demonstrates the transient designer function
%
% See also TRANSIENTDESIGNER, TRANSIENTANALYSIS
clear; clc;
```

```
[in,Fs]=audioread('AcGtr.wav');

% Attack and sustain parameters [-1,+1]
attack = 0;
sustain = 1;

out = transientDesigner(in,attack,sustain);

plot(out,'r'); hold on;
plot(in,'b'); hold off;
legend({'Output','Input'},'FontSize',14);

sound(out,Fs);
```

Example 17.9: Script: `transientExample.m`

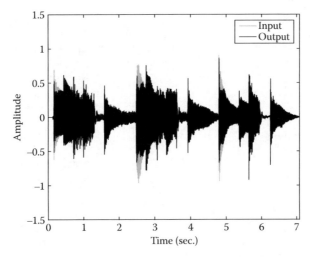

Figure 17.20: Waveform of signal processed by transient designer effect to decrease attack

```
% TRANSIENTDESIGNER
% This function implements the transient designer
% audio effect. First, a detection analysis is performed
% to determine the sections of the signal that should
% be labeled "attack" and "sustain". Then the amplitude
```

```
% of these sections is scaled based on the input parameters
%
% Input Variables
%    attack : amount to change transient (-1 dec, 0 unity, +1 inc)
%    sustain : amount to change sustain
%
% See also TRANSIENTANALYSIS


function [out] = transientDesigner(in,attack,sustain)

N = length(in);
% Initialize filtering parameters
gFast = 0.9991;    % Feedback gain for the "fast" envelope
fbFast = 0;        % Variable used to store previous envelope value
gSlow = 0.9999;    % Feedback gain for "slow" envelope
fbSlow = 0;
envFast = zeros(N,1);
envSlow = zeros(N,1);
differenceEnv = zeros(N,1);

% Measure fast and slow envelopes
for n = 1:N

    envFast(n,1) = (1-gFast) * 2 * abs(in(n,1)) + gFast * fbFast;
    fbFast = envFast(n,1);


    envSlow(n,1) = (1-gSlow) * 3 * abs(in(n,1)) + gSlow * fbSlow;
    fbSlow = envSlow(n,1);

    % Create the difference envelope between "fast" and "slow"
    differenceEnv(n,1) = envFast(n,1) - envSlow(n,1);
    % Note: difference envelope will have a positive value
    % when envFast is greater than envSlow. This occurs
    % when the signal is in "attack". If the difference
    % envelope is negative, then the signal is in
    % "sustain".
end

attEnv = zeros(N,1);
```

```
susEnv = zeros(N,1);
% Separate attack and sustain envelopes
for n = 1:N

    if differenceEnv(n,1) > 0 % "Attack" section

        attEnv(n,1) = (attack * differenceEnv(n,1)) + 1;
        susEnv(n,1) = 1; % No change

    else % "Sustain" section

        attEnv(n,1) = 1; % No change
        susEnv(n,1) = (sustain * -differenceEnv(n,1)) + 1;

    end

end

% Apply the attack and sustain envelopes
out = (in .* attEnv) .* susEnv;
```

Example 17.10: Function: `transientDesigner.m`

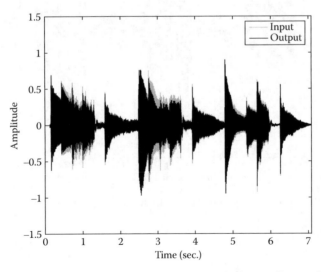

Figure 17.21: Waveform of signal processed by transient designer effect to decrease sustain

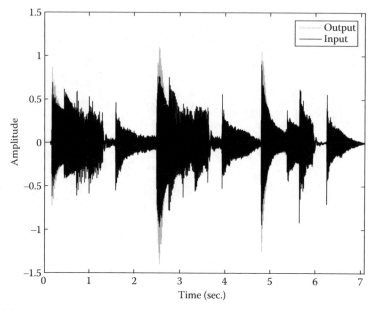

Figure 17.22: Waveform of signal processed by transient designer effect to increase attack

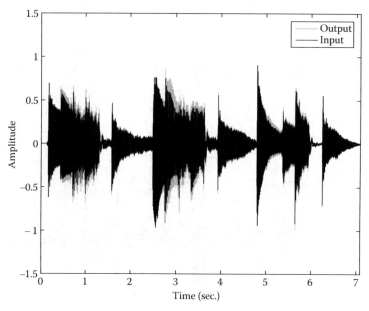

Figure 17.23: Waveform of signal processed by transient designer effect to increase sustain

Bibliography

Shannon, R. V., Zeng, F. G., Kamath, V., Wygonski, J., & Ekelid, M. (1995). Speech recongnition with primarily temporal cues. *Science, 270*, 303–304.

Chapter 18

Dynamic Range Processors
Compressors and Expanders

18.1 Introduction: Dynamic Range Processors

This chapter introduces another category of audio effects. These processors change the amplitude of a signal, influencing the signal's dynamic range. By definition, dynamic range is the difference between the maximum and minimum amplitude of a signal or system.

Examples of **dynamic range (DR) processors** include compressors and expanders. Additionally, a specific type of compressor, with a distinguishing name, is the limiter. Similarly, a gate is a specific type of expander.

18.1.1 Basics of Amplitude Processing

At a fundamental level, each dynamic range processor is an amplitude processor with additional features. Therefore, the actual processing of the input signal is based on multiplication by a linear scaling value. In Section 6.3, a block diagram is shown for changing the amplitude of the signal, and is repeated again in Figure 18.1.

For amplitude envelope effects in Chapter 17, the modulating gain is a function of sample number, n. Therefore, element-wise multiplication is used for processing, similar to conventional amplitude modulation in Chapter 8. In this chapter, DR processors use a linear control amplitude, $g_{lin}[n]$, which can change over time relative to the amplitude envelope of the input signal.

18.1.2 Compressors: Reduce Dynamic Range

One type of DR processor is the **compressor** (Giannoulis, Massberg, & Reiss, 2012). In theory, a compressor reduces, or compresses, a signal's dynamic range. In practice, this result depends on several parameters of the processor, as well as the input signal. A basic explanation of the processing of a compressor involves reducing the amplitude of a signal

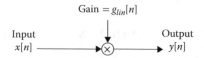

Figure 18.1: Block diagram of basic amplitude processing

when it is greater than a specified level, but allowing the signal to pass through unchanged when the amplitude is below the specified level.

Audio engineers use compressors for several different reasons including to prevent signal clipping, to maintain a consistent signal level, and to shape the ADSR envelope (Section 17.2.2).

18.1.3 Expanders: Increase Dynamic Range

Another type of DR processor is the **expander**. In many ways, the performance of an expander complements the performance of a compressor. In theory, an expander increases, or expands, a signal's dynamic range. Just as with a compressor, this result depends on several factors. Expanders process a signal by reducing its amplitude when the amplitude is less than a specified level, but do not change the amplitude when it is above a specified level.

Expanders are commonly used to remove background noise, or bleed, in a recorded signal. As an example, in a multi-track drum recording, it can be beneficial to reduce the amplitude of the bass drum in the signal captured by the microphone above the snare drum. Additionally, expanders can be used to shape the ADSR envelope of a signal.

18.2 Detection Path

One basic design for a DR processor involves routing a signal through two parallel paths, where one path is responsible for detecting the incoming signal's amplitude. The result of the detection process is a sequence of linear amplitude values that can be multiplied by the corresponding samples of the input signal. When the processor does not change the amplitude of the input signal, $g_{lin}[n] = 1$. When the processor is reducing the amplitude of the input signal, $g_{lin}[n] < 1$. An example of a block diagram with a feedforward detection path is shown in Figure 18.2.

The process of detection, to convert the incoming signal to a sequence of amplitude values, requires several steps. Important parameters to the detection path are the effect's attack, release, threshold (T), ratio (R), and knee width (W). The following sections describe the sequence of analyses as part of the detection path, shown in detail in Figure 18.3.

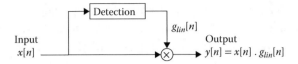

Figure 18.2: Block diagram of a basic DR processor

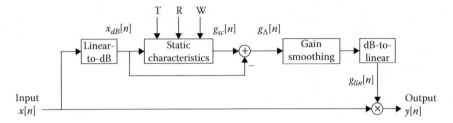

Figure 18.3: Block diagram of a DR processor with a detailed description of the detection path

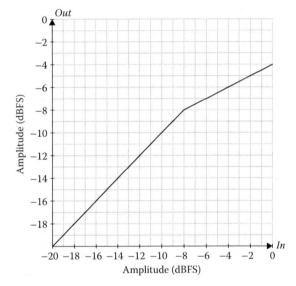

Figure 18.4: Characteristic curve of a compressor

18.3 Conversion to Decibel Scale

DR processors conventionally detect a signal's amplitude relative to a decibel scale. Therefore, the linear amplitude of the input signal in the detection path is first converted to a decibel amplitude. The details of this process are described in Section 17.3.2.

A diagram to describe the characteristic curve of a compressor is shown in Figure 18.4. It represents the relationship between input and output amplitude on a decibel scale. For a digital compressor, the scale is relative to full-scale (FS) amplitude. On a linear scale, the maximum signal magnitude of 1 is defined as the FS reference. Therefore, any linear amplitude, a_{lin}, can be converted to a full-scale decibel (dBFS) amplitude, a_{dBFS}, using $a_{dBFS} = 20 \cdot log_{10} \left(\frac{|a_{lin}|}{1 FS} \right)$.

The amplitude of the decibel, unipolar signal can then be analyzed relative to the decibel scale shown in Figure 18.4. For a digital DR processor, the signal amplitude is within the range of $-\infty$ to 0 dBFS. In practice, it is only necessary to use a minimum level of -144 or -96 dBFS depending on the bit depth of the audio signal. Additionally, values of $-\infty$ can create issues when performing subsequent computational analyses.

Examples: Execute the following commands in MATLAB.

```
% This script demonstrates the process of converting
% the linear amplitude of a signal to a decibel scale.
% An additional step is included to prevent values
% of negative infinity from happening
x = [-1:.01:1].';      % Linear amplitude values over FS range
N = length(x);
for n = 1:N
    x_dB(n,1) = 20 * log10(abs(x(n,1))); % Convert to dB
    if x_dB(n,1) < -144      % Conditional to prevent
        x_dB(n,1) = -144;    % values of negative infinity
    end                      % or anything below noise floor
end
plot(x_dB);
% This plot shows the result of the initial block in the detection
% path converting a linear amplitude to a decibel amplitude
```

Example 18.1: Script: dBConversion.m

18.4 Static Characteristics

After converting the amplitude of the incoming signal to a decibel scale, the next stage is to determine a gain value based on the processor's static characteristics, independent of the processor's attack and release response times. A gain value from the static characteristics, $g_{sc}[n]$, is computed based on several parameters of the processor including threshold, ratio, and knee width.

An example diagram of a compressor's characteristic curve is shown in Figure 18.5, along with the threshold and ratio parameters.

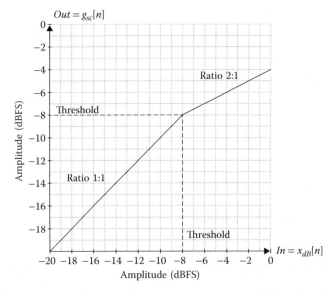

Figure 18.5: Characteristic curve of a compressor with threshold and ratio parameters

18.4.1 Threshold

For a DR processor, the **threshold** is the set amplitude level where the mode of processing changes. For a compressor, no gain reduction is applied when the incoming amplitude level is *less* than the threshold. However, when the incoming amplitude level is *greater* than the threshold, some amount of gain reduction is applied. From the example in Figure 18.5, the threshold is −8 dBFS.

After a threshold has been set, the unipolar signal on the decibel scale can be analyzed to determine when the signal amplitude is greater than, or less than, the threshold. An example to visualize the analysis using the sine wave signal from Figure 17.9 and a threshold of −8 dBFS is shown in Figure 18.6.

18.4.2 Ratio

For a DR processor, the **ratio** is a parameter used in determining the amount of gain reduction applied to the input signal. For a compressor, it defines the necessary amplitude of the input signal in relation to the threshold for the amplitude of the output signal to be 1 dB greater than the threshold. As an example, a ratio of 8:1 means if the amplitude of the input signal is 8 dB greater than the threshold, then the amplitude of the output signal is 1 dB greater than the threshold. Furthermore, with the same ratio, if the amplitude of the input signal is 16 dB greater than the threshold, then the amplitude of the output signal is 2 dB greater than the threshold.

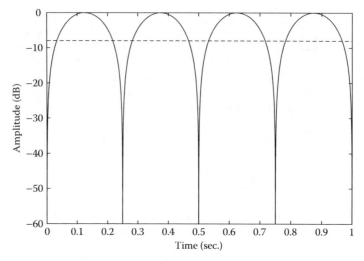

Figure 18.6: Comparing a unipolar signal to the threshold of a DR processor

Conceptually, the ratio changes the relative amount of gain reduction. Assuming the same input signal and threshold, when the ratio is larger, more gain reduction is applied. Ratio values can range from 1:1 for no gain reduction, to ∞:1 for limiting.

Based on Figure 18.4, the desired output amplitude, $g_{sc}[n]$, can be calculated for an input signal, $x_{dB}[n]$, greater than threshold T based on ratio R using the following relationship:

$$g_{sc}[n] = T + \frac{x_{dB}[n] - T}{R}$$

Examples: Execute the following commands in MATLAB.

```
% This script demonstrates how to use the static
% characteristics of a compressor as part of
% the detection path. At the end of the script,
% the characteristic curve is plotted for the
% static characteristics.
x = [0:.001:1].';    % Simple input signal
N = length(x);
% Initialize static characteristics
T = -12;    % Threshold (dBFS)
R = 4;      % Ratio (4:1)
for n = 1:N
```

```
    x_dB(n,1) = 20 * log10(abs(x(n,1)));
    if x_dB(n,1) < -144
        x_dB(n,1) = -144;
    end


    % Comparison to threshold
    if x_dB(n,1) > T
        % Perform compression
        g_sc(n,1) = T + ((x_dB(n,1) - T)/R);
    else
        % Do not compress
        g_sc(n,1) = x_dB(n,1);
    end
end
plot(x_dB,g_sc); % Compressor characteristic curve plot
xlabel('Input Amplitude (dBFS)');
ylabel('Output Amplitude (dBFS)');
```

Example 18.2: Script: `staticCharacteristics.m`

18.4.3 Knee

Many DR processors include a parameter called the **knee**, which influences the calculation of the static characteristics. The knee impacts the transition from the unprocessed, 1:1 ratio to the ratio of gain reduction near the threshold. A soft knee has a gradual transition from one operating level to the other. A hard knee has an immediate transition at the threshold from one operating level to the other.

A variable knee parameter can be defined by the width in decibels. Therefore, the desired gain computer, $g_{sc}[n]$, can be calculated for an input signal, $x_{dB}[n]$, greater than threshold T based on ratio R and width W using the following relationship:

$$g_{sc}[n] = \begin{cases} x_{dB}[n] & \text{if } x_{dB}[n] < \left(T - \frac{W}{2}\right) \\[2ex] x_{dB}[n] + \dfrac{\left(\frac{1}{R}-1\right)\left(x_{dB}[n]-T+\frac{W}{2}\right)^2}{2\cdot W} & \text{if } \left(T - \frac{W}{2}\right) \le x_{dB}[n] \le \left(T + \frac{W}{2}\right) \\[2ex] T + \dfrac{x_{dB}[n]-T}{R} & \text{if } x_{dB}[n] > \left(T + \frac{W}{2}\right) \end{cases}$$

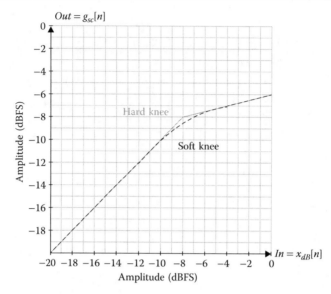

Figure 18.7: Characteristic curve of a soft- and hard-knee compressor

Therefore, a hard-knee compressor is created when the knee width is 0. A visual comparison of a soft-knee and hard-knee compressor is shown in Figure 18.7 with a threshold of −8 dB, ratio of 4:1, and knee width of 4.

Examples: Execute the following commands in MATLAB.

```
% This script demonstrates how to incorporate
% the knee parameter to create a soft-knee compressor.
% At the end of the script, the characteristic curve
% is plotted for the static characteristics.
x = [0:.001:1].';    % Simple input signal
N = length(x);


T = -12;
R = 4;
W = 4;   % Knee width, 4 dB
for n = 1:N
    x_dB(n,1) = 20 * log10(abs(x(n,1)));
    if x_dB(n,1) < -144
        x_dB(n,1) = -144;
    end


    % Comparison to threshold
```

```
    if x_dB(n,1) > (T + W/2)
        % Above knee curve
        g_sc(n,1) = T + ((x_dB(n,1) - T)/R);
    elseif  x_dB(n,1) > (T - W/2)
        % Within knee curve
        g_sc(n,1) = x_dB(n,1) + ...
        ((1/R - 1)*(x_dB(n,1) - T + W/2)^2)/(2*W);
    else
        % Do not compress
        g_sc(n,1) = x_dB(n,1);
    end
end
plot(x_dB,g_sc);   % Compressor characteristic curve plot
xlabel('Input Amplitude (dBFS)');
ylabel('Output Amplitude (dBFS)');
```

Example 18.3: Script: `softKnee.m`

18.4.4 Gain Change Amount

The gain value based on the static characteristics conceptually represents what the theoretical output amplitude should be. However, it does not represent how the input signal should be *changed* to achieve the output amplitude. Therefore, an additional step of processing is included in the detection path to determine the appropriate gain change amount to be applied to the input signal. The gain change amount is determined by comparing the static characteristic gain value to the amplitude of the input signal, $g_\Delta[n] = g_{sc}[n] - x_{dB}[n]$.

Another way to consider the difference between what the amplitude should be, $g_{sc}[n]$, and what it is currently, $x_{dB}[n]$, is to describe the gain change amount, $g_\Delta[n]$, as the *error signal*. The system can then be analyzed, or designed, based on how it responds to the error signal. Therefore, concepts from the field of Control Theory can be used to analyze and implement DR processors.

18.5 Response Time

Up to this point in the detection path, each sample of the input signal has been analyzed independently. In other words, the detection path has performed an element-wise, memoryless analysis of the input signal. If the gain change amount, $g_\Delta[n]$, were simply converted to the linear control amplitude, $g_{lin}[n]$, then the processor would create a hard-clipping distortion

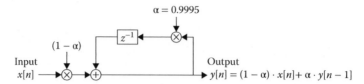

Figure 18.8: Block diagram of system for step response example

effect based on the decibel scale. This would be undesirable, and would not perform the purpose of the effect.

Instead, DR processors are designed so they do not respond instantaneously. The gain change amount, $g_\Delta[n]$, is smoothed to create a gradually changing linear control amplitude, $g_{lin}[n]$. The smoothing is performed so the system responds based on the attack and release parameters of the DR processor.

18.5.1 Step Response Analysis

The method used to evaluate the response time of a DR processor is called the **step response**. It is as fundamental and important to analyzing DR processors as the impulse response (Section 11.7) is to analyzing echo, reverb, and filter effects. With the step response, a system is analyzed for how it transitions from one state to another.

For a compressor, a step response is used to analyze how the effect transitions from a state of noncompression (when the signal starts below the threshold) to a state of compression (when the signal goes above the threshold). This is called the attack of the compressor. Additionally, the step response is used to analyze the release of the compressor. This occurs when the effect transitions from a state of compression (when the signal starts above the threshold) to a state of noncompression (when the signal drops below the threshold).

The input signal used to measure the step response of a system is a signal that begins with an amplitude value of 0. After some number of samples, the amplitude of the input signal immediately changes to a value of 1 for the remaining samples, where the amplitude value is held constant.

As an example, consider the step response of the system shown in Figure 18.8. The measurement of the system is conducted in Example 18.4.

Examples: Create and run the following m-file in MATLAB.

```
% STEPDEMO
% This script demonstrates the process of measuring
% the step response of a first-order, feedback LPF.
```

```
% A plot is created showing a comparison between
% the input step signal and the output response.

% Initialize the sampling rate
Fs = 48000; Ts = 1/Fs;

% Create step input signal
x = [zeros(Fs,1) ; ones(Fs,1)];
N = length(x);
% Initialize gain value
alpha = 0.9995;  % Also try values between 0.999-0.9999
q = 0; % Initialize feedback variable
for n = 1:N
    y(n,1) = (1-alpha) * x(n,1) + alpha * q;
    q = y(n,1); % Stores the "previous" value for the next loop cycle
end

t = [0:N-1]*Ts; % Time vector for plot
plot(t,x,t,y);
axis([0 2 -0.1 1.1]); xlabel('Time (sec.)');
legend('Step Input','Output');
```

Example 18.4: Script: `stepDemo.m`

In Figure 18.9, the output signal does not immediately transition from a value of 0 to a value of 1 an amplitude of 1, unlike the input signal. Rather, the output signal gradually transitions as a response to the step input.

18.5.2 *MATLAB* `stepz` *Function*

A built-in MATLAB function can be used to analyze the step response of a sampled, discrete-time system. The following syntax can be used to call the function: `h = stepz(b,a)`. This function finds the step response, `h`, of the system represented by arrays `b` and `a`. Following the same convention as MATLAB filter functions, the coefficients for feedforward paths are placed in array `b` and the coefficients for feedback paths are place in array `a`. Therefore, the system in Figure 18.8 can be represented by arrays `b = [1-alpha]` and `a = [1 , -alpha]`. In Example 18.5, the `stepz` function is demonstrated, along with the result of varying the value of the feedback gain, α. By varying the feedback gain, the system's step response takes a different amount of time to transition from one state to another (Figure 18.10).

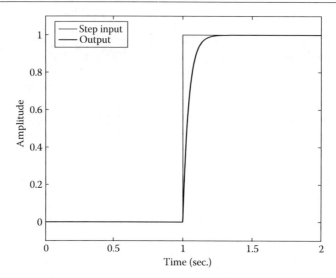

Figure 18.9: Step response of example system

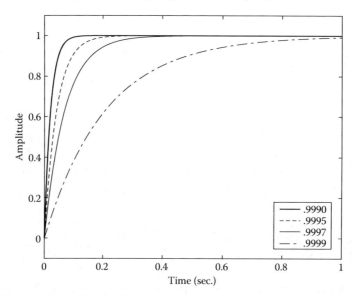

Figure 18.10: Varying the feedback gain in the system step response

Examples: Create and run the following m-file in MATLAB.

```
% STEPRESPONSE
% This script demonstrates the built-in
% MATLAB function stepz(b,a). The step
% response for several first-order systems
% is compared using different feedback gains.
```

```
%
% See also STEPZ


Fs = 48000; % Initialize the sampling rate
sec = 1;   % Time length in seconds
n = sec*Fs; % Convert seconds to # of samples


% Define different gain values to test
gains = [0.999;0.9995;0.9997;0.9999];


% Determine a new step response each time through the loop
for element = 1:length(gains)
    alpha = gains(element,1);
    b = [(1-alpha)];
    a = [1 , -alpha];
    [h,t] = stepz(b,a,n,Fs);
    plot(t,h);hold on;
end


hold off; axis([0 sec -0.1 1.1]); xlabel('Time (sec.)');
AX = legend('.9990','.9995','.9997','.9999');set(AX,'FontSize',12);
```

Example 18.5: Script: stepResponse.m

18.5.3 *System Response Due to Step Input*

For the system shown in Figure 18.9 and plotted in Figure 18.10, the response due to step input, $s[n]$, can be represented by the following equation:

$$s[n] = 1 - \alpha \cdot \alpha^n \tag{18.1}$$

where n is the sample number. Using the sampling period, T_s, this expression can also be represented for time units of seconds where $t = (n-1) \cdot T_s$ using MATLAB array indexing. This relationship can be rearranged to solve for the sample number: $n = \frac{t}{T_s} + 1$. Substituting into Equation (18.1), the response due to step input, $s[t]$, for units of seconds is:

$$s[t] = 1 - \alpha \cdot \alpha^{\left(\frac{t}{T_s}+1\right)}$$

$$s[t] = 1 - \alpha \cdot \alpha^{\frac{t}{T_s}} \cdot \alpha$$

$$s[t] = 1 - \alpha^2 \cdot \alpha^{\frac{t}{T_s}}$$

18.5.4 Rise Time of Step Response

The **rise time** (used for attack and release time of a DR processor) is defined as the time of the step response to transition from 10% to 90% of the final value. A unit step signal has a final value of 1. Therefore, the time, t_{10}, the step response reaches 10% of the final value is defined by $s[t_{10}] = 0.1$. Similarly, the time, t_{90}, the step response reaches 90% of the final value is defined by $s[t_{90}] = 0.9$. Used together, the rise time is $t_r = t_{90} - t_{10}$.

The process to determine the rise time of the step response for a system with feedback gain α is the following. First, time t_{10} is determined.

$$s[t_{10}] = 0.1 = 1 - \alpha^2 \cdot \alpha^{\frac{t_{10}}{T_s}}$$

$$\alpha^2 \cdot \alpha^{\frac{t_{10}}{T_s}} = 0.9$$

$$\alpha^{\frac{t_{10}}{T_s}} = \frac{0.9}{\alpha^2}$$

$$ln\left(\alpha^{\frac{t_{10}}{T_s}}\right) = ln\left(\frac{0.9}{\alpha^2}\right)$$

$$\frac{t_{10}}{T_s} \cdot ln(\alpha) = ln\left(\frac{0.9}{\alpha^2}\right)$$

$$t_{10} = ln\left(\frac{0.9}{\alpha^2}\right) \cdot \frac{T_s}{ln(\alpha)}$$

Similarly, time t_{90} can be found:

$$s[t_{90}] = 0.9 = 1 - \alpha^2 \cdot \alpha^{\frac{t_{90}}{T_s}}$$

$$\alpha^2 \cdot \alpha^{\frac{t_{90}}{T_s}} = 0.1$$

$$\alpha^{\frac{t_{90}}{T_s}} = \frac{0.1}{\alpha^2}$$

$$ln\left(\alpha^{\frac{t_{90}}{T_s}}\right) = ln\left(\frac{0.1}{\alpha^2}\right)$$

$$\frac{t_{90}}{T_s} \cdot ln(\alpha) = ln\left(\frac{0.1}{\alpha^2}\right)$$

$$t_{90} = ln\left(\frac{0.1}{\alpha^2}\right) \cdot \frac{T_s}{ln(\alpha)}$$

This leads to the result for rise time: $t_r = t_{90} - t_{10}$.

$$t_r = ln\left(\frac{0.1}{\alpha^2}\right) \cdot \frac{T_s}{ln(\alpha)} - ln\left(\frac{0.9}{\alpha^2}\right) \cdot \frac{T_s}{ln(\alpha)}$$

$$t_r = \frac{T_s}{ln(\alpha)} \cdot \left[ln\left(\frac{0.1}{\alpha^2}\right) - ln\left(\frac{0.9}{\alpha^2}\right)\right]$$

$$t_r = \frac{T_s}{ln(\alpha)} \cdot \left[\left(ln(0.1) - ln(\alpha^2)\right) - \left(ln(0.9) - ln(\alpha^2)\right)\right]$$

$$t_r = \frac{T_s}{ln(\alpha)} \cdot \left[ln(0.1) - ln(0.9)\right]$$

$$t_r = \frac{T_s}{-ln(\alpha)} \cdot \left[ln(0.9) - ln(0.1)\right]$$

$$t_r = \frac{T_s}{-ln(\alpha)} \cdot \left[ln(\frac{0.9}{0.1})\right]$$

$$t_r = \frac{T_s}{-ln(\alpha)} \cdot \left[ln(9)\right]$$

If a desired rise time (i.e., attack time or release time) is known, the corresponding α value for Figure 18.10 can be determined using the following rearrangement:

$$t_r = \frac{T_s \cdot ln(9)}{-ln(\alpha)}$$

$$-ln(\alpha) = \frac{T_s \cdot ln(9)}{t_r}$$

$$ln(\alpha) = \frac{-ln(9)}{F_s \cdot t_r}$$

$$\alpha = e^{\frac{-ln(9)}{F_s \cdot t_r}}$$

18.5.5 Using Response Time for a Compressor

This same approach can be used to change the response time of a DR processor. The system in Figure 18.8 can be included in the detection path to smooth between subsequent gain change values, $g_\Delta[n]$. By adjusting the feedback gain, α, the attack time and release time of the DR processor are changed.

Examples: Create and run the following m-file in MATLAB.

```
% BASICCOMP
% This script creates a dynamic range compressor with
% attack and release times linked together. A step
```

```
% input signal is synthesized for testing. A plot
% is produced at the end of the script to show
% a comparison of the input step signal, the output
% response, and the gain reduction curve.
%
% See also COMPRESSOR, COMPRESSOREXAMPLE

% Step input signal
Fs = 48000; Ts = 1/Fs;
x = [zeros(Fs,1); ones(Fs,1); zeros(Fs,1)];
N = length(x);
% Parameters for compressor
T = -12;    % Threshold = -12 dBFS
R = 3;      % Ratio = 3:1
responseTime = 0.25;  % time in seconds
alpha = exp(-log(9)/(Fs * responseTime));
gainSmoothPrev = 0; % Initialize smoothing variable

y = zeros(N,1);
lin_A = zeros(N,1);
% Loop over each sample to see if it is above thresh
for n = 1:N
    %%%%% Calculations of the detection path
    % Turn the input signal into a unipolar signal on the dB scale
    x_uni = abs(x(n,1));
    x_dB = 20*log10(x_uni/1);
    % Ensure there are no values of negative infinity
    if x_dB < -96
        x_dB = -96;
    end

    % Static characteristics
    if x_dB > T
        gainSC = T + (x_dB - T)/R; % Perform downwards compression
    else
        gainSC = x_dB; % Do not perform compression
    end

    gainChange_dB = gainSC - x_dB;
```

```
    % smooth over the gainChange_dB to alter response time
    gainSmooth =  ((1-alpha)*gainChange_dB) + (alpha*gainSmoothPrev);

    % Convert to linear amplitude scalar
    lin_A(n,1) = 10^(gainSmooth/20);

    %%%%%% Apply linear amplitude from detection path
    %%%%%% to input sample
    y(n,1) = lin_A(n,1) * x(n,1);

    % Update gainSmoothPrev used in the next sample of the loop
    gainSmoothPrev = gainSmooth;
end
t = [0:N-1]*Ts; t = t(:);


subplot(3,1,1);
plot(t,x); title('Step Input');axis([0 3 -0.1 1.1]);
subplot(3,1,2);
plot(t,y); title('Comp Out'); axis([0 3 -0.1 1.1]);
subplot(3,1,3);
plot(t,lin_A); title('Gain Reduction');axis([0 3 -0.1 1.1]);
% The "gain reduction" line shows the amount of compression
% applied at each sample of the signal. When the value
% is "1", there is no compression. When the value is less
% than "1", gain reduction is happening.
```

Example 18.6: Script: basicComp.m

As can be seen in Figure 18.11, the gain reduction gradually transitions when the input amplitude increases from 0 to 1, as well as when the input amplitude decreases from 1 to 0. These two cases represent the response of a compressor's attack and release.

18.5.6 *Attack and Release Time*

The **attack time** for a DR processor represents the speed at which it responds to the amplitude of an input signal initially crossing the threshold. For a compressor, attack time is the response rate while the signal amplitude is increasing and above the threshold. For an expander, attack time is the response rate while the signal amplitude is increasing and below the threshold.

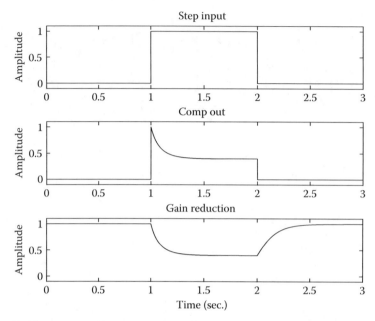

Figure 18.11: Input amplitude, output amplitude, and gain reduction of a compressor

The **release time** for a DR processor represents the speed at which it returns to the state where it is not changing the amplitude of the input signal. For a compressor, release time is the response rate while the signal amplitude is decreasing after the signal amplitude is greater than the threshold. For an expander, release time is the response rate while the signal amplitude is decreasing after the signal amplitude is less than the threshold.

It is common for the response time of DR processors to be different in attack and release modes. This way, an engineer may set separate controls for the attack and release time. Therefore, it is necessary to differentiate when a DR processor should be in attack mode and when it should be in release mode.

A compressor is in attack mode under the following conditions. When the incoming signal's amplitude, x_{dB}, is greater than the threshold, the gain based on the static characteristics, g_{sc}, is less than x_{dB} (see Figure 18.4). Therefore, the gain change amount, $g_\Delta = g_{sc} - x_{dB}$, is negative. Finally, it is necessary for g_Δ to be less than the previous gain smoothed amount (initially set to 0 dB) for the compressor to be in attack mode.

As a consequence, the compressor is in a release mode while g_Δ is greater than the previous gain smoothed amount until it returns to zero gain reduction or it switches back to attack mode.

In Example 18.8, an implementation of a feedforward compressor is demonstrated. The result of the processing is shown in Figure 18.12.

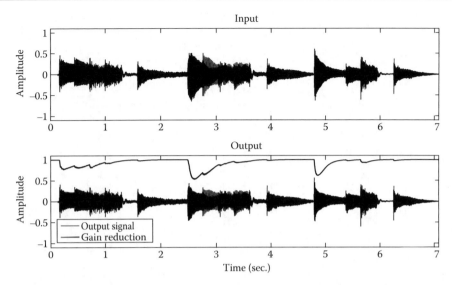

Figure 18.12: Waveform and gain reduction for signal processed by a feedfoward compressor

Examples: Create and run the following m-files in MATLAB.

```
% COMPRESSOREXAMPLE
% This script creates a dynamic range compressor with
% separate attack and release times
%
% See also COMPRESSOR, BASICCOMP

% Acoustic guitar "audio" sound file
[in,Fs] = audioread('AcGtr.wav');

% Parameters for compressor
T = -15;    % Threshold = -15 dBFS
R = 10;     % Ratio = 10:1

% Initialize separate attack and release times
attackTime = 0.05;  % time in seconds
releaseTime = 0.25; % time in seconds

% Compressor function
out = compressor(in,Fs,T,R,attackTime,releaseTime);
sound(out,Fs);
```

Example 18.7: Script: `compressorExample.m`

```
% COMPRESSOR
% This function implements a dynamic range compressor
% with separate attack and release times. To visualize
% the results, code is included at the end of the function
% to plot the input signal and the compressed output signal.
% As a default, this code is commented so it does not create
% a plot each time the function is called.
%
% Input Variables
%   T : threshold relative to 0 dBFS
%   R : ratio (R to 1)
%   attackTime : units of seconds
%   releaseTime : units of seconds
%
% See also COMPRESSOREXAMPLE, BASICCOMP

function [y] = compressor(x,Fs,T,R,attackTime,releaseTime)
N = length(x);
y = zeros(N,1);
lin_A = zeros(N,1);

% Initialize separate attack and release times
alphaA = exp(-log(9)/(Fs * attackTime));
alphaR = exp(-log(9)/(Fs * releaseTime));

gainSmoothPrev = 0; % Initialize smoothing variable

% Loop over each sample to see if it is above thresh
for n = 1:N
    % Turn the input signal into a unipolar signal on the dB scale
    x_uni = abs(x(n,1));
    x_dB = 20*log10(x_uni/1);
    % Ensure there are no values of negative infinity
    if x_dB < -96
        x_dB = -96;
    end

    % Static characteristics
```

```
        if x_dB > T
            gainSC = T + (x_dB - T)/R; % Perform downwards compression
        else
            gainSC = x_dB; % Do not perform compression
        end


        gainChange_dB = gainSC - x_dB;


        % Smooth over the gainChange
        if gainChange_dB < gainSmoothPrev
            % attack mode
            gainSmooth = ((1-alphaA)*gainChange_dB) ...
                +(alphaA*gainSmoothPrev);
        else
            % release
            gainSmooth = ((1-alphaR)*gainChange_dB) ...
                +(alphaR*gainSmoothPrev);
        end


        % Convert to linear amplitude scalar
        lin_A(n,1) = 10^(gainSmooth/20);


        % Apply linear amplitude to input sample
        y(n,1) = lin_A(n,1) * x(n,1);


        % Update gainSmoothPrev used in the next sample of the loop
        gainSmoothPrev = gainSmooth;
end

% Uncomment for visualization
% t = [0:N-1]/Fs; t = t(:);
% subplot(2,1,1);
% plot(t,x); title('Input');axis([0 t(end) -1.1 1.1]);
% subplot(2,1,2);
% plot(t,y,t,lin_A); title('Output'); axis([0 t(end) -1.1 1.1]);
% legend('Output Signal','Gain Reduction');
```

Example 18.8: Function: `compressor.m`

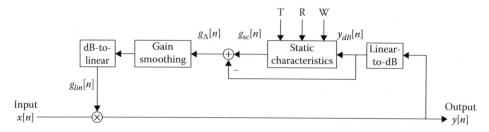

Figure 18.13: Block diagram with the detection path for a feedback compressor

18.6 Additional Detection Path Methods

There are several other types of compressors that can be created by modifying the detection path. These methods include feedback detection, root-mean-square (RMS) detection, and side-chain detection.

18.6.1 Feedback Detection

In the DR processor previously presented, the detection path was based on a feedforward design, where the input signal fed the detection analysis block (Figure 18.2). A different approach would be to use the output signal to create **feedback detection**. A block diagram for this type of processor is shown in Figure 18.13.

As a rule of thumb, a feedback compressor produces a smoother, less dramatic effect. This is a result of g_{lin} calculated based on amplitude values that may have already been slightly compressed. Therefore, the gain change, g_Δ, may not always be as large as the feedforward design.

In Example 18.9, an implementation of a feedback compressor is demonstrated. The result of the processing is shown in Figure 18.14.

Examples: Create and run the following m-file in MATLAB.

```
% FEEDBACKCOMP
% This script creates a feedback compressor.
% The processing of the detection path is similar
% to the feedforward compressor. The main difference
% is the output "y" is analyzed in the detection
% path, not the input "x". A plot is produced
% at the end of the script to visualize the result.
```

```
%
% See also COMPRESSOR, BASICCOMP


% Acoustic guitar "audio" sound file
[x,Fs] = audioread('AcGtr.wav');


% Parameters for compressor
T = -15;    % Threshold = -15 dBFS
R = 10;     % Ratio = 10:1


% Initialize separate attack and release times
attackTime = 0.05;  % time in seconds
alphaA = exp(-log(9)/(Fs * attackTime));
releaseTime = 0.25;  % time in seconds
alphaR = exp(-log(9)/(Fs * releaseTime));


gainSmoothPrev = 0; % Initialize smoothing variable


y_prev = 0; % Initialize output for feedback detection


N = length(x);
y = zeros(N,1);
lin_A = zeros(N,1);
% Loop over each sample to see if it is above thresh
for n = 1:N
    %%%% Detection path based on the output signal, not "x"
    % Turn the input signal into a unipolar signal on the dB scale
    y_uni = abs(y_prev);
    y_dB = 20*log10(y_uni/1);

    % Ensure there are no values of negative infinity
    if y_dB < -96
        y_dB = -96;
    end

    % Static characteristics
    if y_dB > T
        gainSC = T + (y_dB - T)/R; % Perform downwards compression
    else
        gainSC = y_dB; % Do not perform compression
```

```
        end

        gainChange_dB = gainSC - y_dB;

        % smooth over the gainChange
        if gainChange_dB < gainSmoothPrev
            % attack mode
            gainSmooth = ((1-alphaA)*gainChange_dB) ...
                +(alphaA*gainSmoothPrev);
        else
            % release
            gainSmooth = ((1-alphaR)*gainChange_dB) ...
                +(alphaR*gainSmoothPrev);
        end

        % Convert to linear amplitude scalar
        lin_A(n,1) = 10^(gainSmooth/20);

        % Apply linear amplitude to input sample
        y(n,1) = lin_A(n,1) * x(n,1);
        y_prev = y(n,1); % Update for next cycle

        % Update gainSmoothPrev used in the next sample of the loop
        gainSmoothPrev = gainSmooth;

end
t = [0:N-1]/Fs; t = t(:);

subplot(2,1,1);
plot(t,x); title('Input Signal');axis([0 7 -1.1 1.1]);
subplot(2,1,2);
plot(t,y,t,lin_A); title('Output'); axis([0 7 -1.1 1.1]);
legend('Output Signal','Gain Reduction');
```

Example 18.9: Script: feedbackComp.m

18.6.2 RMS Detection

Another approach to DR processors uses the RMS amplitude for detecting and processing a signal. Here, the detection is based on a conceptual *average* amplitude rather than the

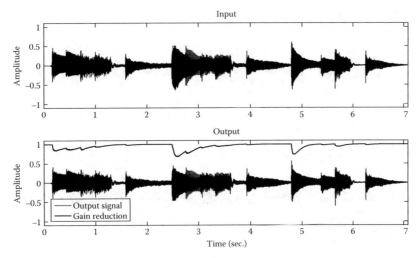

Figure 18.14: Waveform and gain reduction for signal processed by a feedback compressor

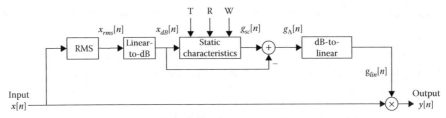

Figure 18.15: Block diagram with the detection path for an RMS compressor

instantaneous, or *peak*, amplitude. The following can be used to calculate a signal's RMS for a DR processor:

$$(x_{rms}[n])^2 = \frac{1}{M} \sum_{m=-\frac{M}{2}}^{\frac{M}{2}-1} (x[n+m])^2$$

The number of included samples, M, for the RMS amplitude can be set to change the length of the time window for the calculation. A longer time window typically results in a smoother compressor, and a shorter time window approaches the performance of the peak detector. To achieve results for a typical DR processor, a time window of 1000+ samples should be used.

The calculation of RMS amplitude can be incoporated in the detection path prior to converting the input amplitude to the decibel scale. Because the RMS value is an approximation of the signal's average strength, it is not necessary to perform gain smoothing in this approach. A block diagram including this additional step is shown in Figure 18.15.

The limitation of this method is controlling the response time. Both the attack and release time are based on the length of the window used in the RMS calculation.

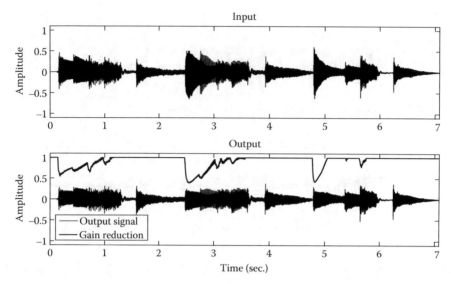

Figure 18.16: Waveform and gain reduction for signal processed by an RMS compressor

In Example 18.10, an implementation of a RMS compressor is demonstrated. The result of the processing is shown in Figure 18.16.

Examples: Create and run the following m-file in MATLAB.

```
% RMSCOMP
% This script creates a compressor with
% conventional RMS detection. The RMS value
% is calculated over a range of "M" samples.
% Note: attack and release are linked
%
% See also COMPRESSOR, RMSCOMP2
clear;clc;close all;

% Acoustic guitar "audio" sound file
[x,Fs] = audioread('AcGtr.wav');

% Parameters for compressor
T = -20;    % Threshold = -20 dBFS
R = 4;      % Ratio = 4:1

% Initialize separate attack and release times
attackTime = 0.1;  % time in seconds
```

Dynamic Range Processors 435

```
alphaA = exp(-log(9)/(Fs * attackTime));
releaseTime = 0.25;  % time in seconds
alphaR = exp(-log(9)/(Fs * releaseTime));


gainSmoothPrev = 0;  % Initialize smoothing variable


M = 2048;  % length of RMS calculation

% Initialize the first time window in a buffer
x_win = [zeros(M/2,1);x(1:(M/2),1)];


N = length(x);
y = zeros(N,1);
lin_A = zeros(N,1);

% Loop over each sample to see if it is above thresh
for n = 1:N

    % Calculate the RMS for the current window
    x_rms = sqrt(mean(x_win.^2));

    % Turn the input signal into a unipolar signal on the dB scale
    x_dB = 20*log10(x_rms);

    % Ensure there are no values of negative infinity
    if x_dB < -96
        x_dB = -96;
    end


    % Static characteristics
    if x_dB > T
        gainSC = T + (x_dB - T)/R;  % Perform compression
    else
        gainSC = x_dB;  % Do not perform compression
    end


    gainChange_dB = gainSC - x_dB;


    % Convert to linear amplitude scalar
    lin_A(n,1) = 10^(gainChange_dB/20);
```

```
    % Apply linear amplitude to input sample
    y(n,1) = lin_A(n,1) * x(n,1);


    % Update the current time window
    if n+(M/2) < N
        x_win = [x_win(2:end,1) ; x(n+(M/2)+1,1)];
    else
        x_win = [x_win(2:end,1) ; 0];
    end
end
t = [0:N-1]/Fs; t = t(:);

subplot(2,1,1);
plot(t,x); title('Input Signal');axis([0 t(end) -1.1 1.1]);
subplot(2,1,2);
plot(t,y,t,lin_A); title('Output'); axis([0 t(end) -1.1 1.1]);
legend('Output Signal','Gain Reduction');
```

Example 18.10: Script: rmsComp.m

Approximating RMS for Real-Time Processors

The calculation of the RMS value over a time window with 1000+ samples for each sample in a signal is computationally complex. Comparable results can be achieved by approximating the RMS value using feedback. In this method, the RMS value is approximated by taking the square root of the current sample squared, plus the previous RMS value. Because of feedback, the previous RMS value is a combination of all previous samples of the signal squared. This creates the following gain smoothing process:

$$y[n] = -\sqrt{(1-\alpha)\cdot(x[n])^2 + \alpha\cdot(y[n-1])^2} \qquad (18.2)$$

The RMS approximation method can be incorporated into the compressor effect in Figure 18.3 by substituting the gain smoothing in Equation (18.2). Separate attack and release times can be implemented, as shown previously in Example 18.6. The result of the processing is shown in Figure 18.17.

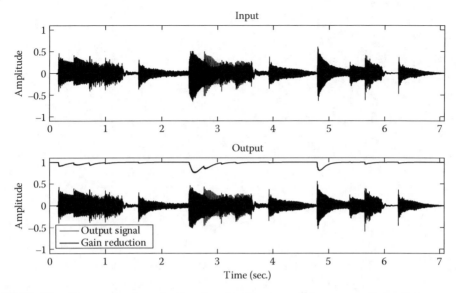

Figure 18.17: Waveform and gain reduction for signal processed by a compressor that uses an RMS approximation

Examples: Create and run the following m-file in MATLAB.

```
% RMSCOMP2
% This script creates a compressor with approximated
% RMS detection. The RMS value is estimated
% by using feedback. Separate attack and release
% times can be achieved.
%
% See also RMSCOMP
clear;clc;close all;

% Acoustic guitar "audio" sound file
[x,Fs] = audioread('AcGtr.wav');

% Parameters for compressor
T = -12;    % Threshold = -12 dBFS
R = 4;      % Ratio = 4:1

% Initialize separate attack and release times
attackTime = 0.1;  % time in seconds
```

```
alphaA = exp(-log(9)/(Fs * attackTime));
releaseTime = 0.1;  % time in seconds
alphaR = exp(-log(9)/(Fs * releaseTime));


gainSmoothPrev = 0; % Initialize smoothing variable


N = length(x);
y = zeros(N,1);
lin_A = zeros(N,1);
% Loop over each sample to see if it is above thresh
for n = 1:N


    % Turn the input signal into a unipolar signal on the dB scale
    x_dB = 20*log10(abs(x(n,1)));


    % Ensure there are no values of negative infinity
    if x_dB < -96
        x_dB = -96;
    end


    % Static characteristics
    if x_dB > T
        gainSC = T + (x_dB - T)/R; % Perform downwards compression
    else
        gainSC = x_dB; % Do not perform compression
    end


    gainChange_dB = gainSC - x_dB;


    % Smooth over the gainChange
    if gainChange_dB < gainSmoothPrev
        % attack mode
        gainSmooth = -sqrt(((1-alphaA)*gainChange_dB^2) ...
            +(alphaA*gainSmoothPrev^2));
    else
        % release
        gainSmooth = -sqrt(((1-alphaR)*gainChange_dB^2) ...
            +(alphaR*gainSmoothPrev^2));
    end
```

```
    % Convert to linear amplitude scalar
    lin_A(n,1) = 10^(gainSmooth/20);

    % Apply linear amplitude to input sample
    y(n,1) = lin_A(n,1) * x(n,1);

    % Update gainSmoothPrev used in the next sample of the loop
    gainSmoothPrev = gainSmooth;

end
t = [0:N-1]/Fs; t = t(:);

subplot(2,1,1);
plot(t,x); title('Input Signal');axis([0 t(end) -1.1 1.1]);
subplot(2,1,2);
plot(t,y,t,lin_A); title('Output'); axis([0 t(end) -1.1 1.1]);
legend('Output Signal','Gain Reduction');
```

Example 18.11: Script: `rmsComp2.m`

18.6.3 Side-Chain Compressor

Rather than simply controlling a signal's dynamic range or shaping the ADSR envelope, compressors can also be used for other musical effects. One interesting method is the **side-chain compressor**. In this method, gain reduction is applied to one signal while the detection path of a compressor is triggered by a different signal.

A common technique in electronic styles of music is to use a kick drum signal to trigger gain reduction on a synthesizer signal. The result is the sythesizer pumps with the rhythm of the song, while also making perceptual space for the kick drum to punch through the mix. A block diagram of this effect is shown in Figure 18.18.

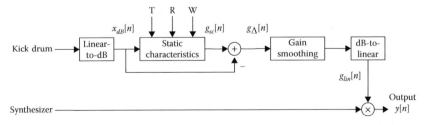

Figure 18.18: Block diagram of a compressor with a detection path for side-chain compression

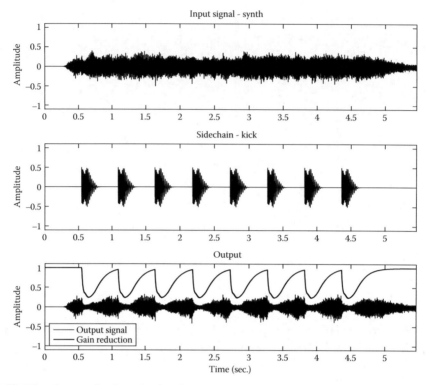

Figure 18.19: Waveform and gain reduction for synthesizer signal processed by a compressor detecting a kick drum in the side chain

In Example 18.12, a side-chain compressor is implemented using sound files of a synthesizer and kick drum included in the book's *Additional Content*. The result of the processing is shown in Figure 18.19.

Examples: Create and run the following m-file in MATLAB.

```
% SIDECHAINCOMP
% This script creates a side-chain compressor with
% a synthesizer signal and kick drum signal.
%
% See also COMPRESSOR
clear;clc;
% Synthesizer input signal
[x,Fs] = audioread('Synth.wav');
% Kick drum for the detection path
[sc] = audioread('Kick.wav');

% Parameters for compressor
```

```
T = -24;    % Threshold = -24 dBFS
R = 10;      % Ratio = 10:1


% Initialize separate attack and release times
attackTime = 0.05;   % time in seconds
alphaA = exp(-log(9)/(Fs * attackTime));
releaseTime = 0.25;   % time in seconds
alphaR = exp(-log(9)/(Fs * releaseTime));


gainSmoothPrev = 0; % Initialize smoothing variable

N = length(sc);
y = zeros(N,1);
lin_A = zeros(N,1);
% Loop over each sample to see if it is above thresh
for n = 1:N
    %%%%% Detection path based on the kick drum input signal
    % Turn the input signal into a unipolar signal on the dB scale
    sc_uni = abs(sc(n,1));
    sc_dB = 20*log10(sc_uni/1);
    % Ensure there are no values of negative infinity
    if sc_dB < -96
        sc_dB = -96;
    end

    % Static characteristics
    if sc_dB > T
        gainSC = T + (sc_dB - T)/R; % Perform downwards compression
    else
        gainSC = sc_dB; % Do not perform compression
    end

    gainChange_dB = gainSC - sc_dB;

    % smooth over the gainChange
    if gainChange_dB < gainSmoothPrev
        % attack mode
        gainSmooth = ((1-alphaA)*gainChange_dB) ...
            +(alphaA*gainSmoothPrev);
    else
```

```
      % release
      gainSmooth = ((1-alphaR)*gainChange_dB) ...
          +(alphaR*gainSmoothPrev);
  end


  % Convert to linear amplitude scalar
  lin_A(n,1) = 10^(gainSmooth/20);


  % Apply linear amplitude to synthesizer input sample
  y(n,1) = lin_A(n,1) * x(n,1);


  % Update gainSmoothPrev used in the next sample of the loop
  gainSmoothPrev = gainSmooth;
end
t = [0:N-1]/Fs; t = t(:);


subplot(3,1,1);
plot(t,x); title('Input Signal - Synth');axis([0 t(end) -1.1 1.1]);
subplot(3,1,2);
plot(t,sc); title('Sidechain - Kick');axis([0 t(end) -1.1 1.1]);
subplot(3,1,3);
plot(t,y,t,lin_A); title('Output'); axis([0 t(end) -1.1 1.1]);
legend('Output Signal','Gain Reduction');


sound(y,Fs);
```

Example 18.12: Script: `sidechainComp.m`

18.7 Expander

An **expander** is a type of DR processor that decreases the amplitude of a signal when it is less than a threshold. When the amplitude of the input signal is greater than the threshold, the signal passes through unprocessed. Therefore, an expander can be created by using a ratio of 1:1 when amplitude is greater than the threshold and a ratio other than 1:1 when amplitude is less than the threshold. For an expander, a ratio of 2:1 decreases the amplitude of the output signal by 2 decibels for every 1 decibel the input signal is below the threshold.

Audio engineers use expanders for several purposes including reducing the amplitude of background noise or bleed in a recording and shaping the ADSR envelope of a signal. A diagram describing the static characteristics of an expander is shown in Figure 18.20.

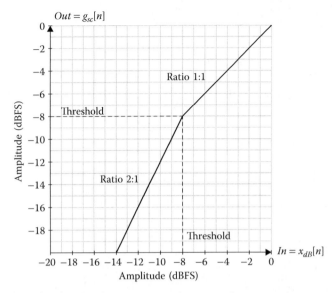

Figure 18.20: Characteristic curve of an expander

Based on Figure 18.20, the desired output amplitude, $g_{sc}[n]$, can be calculated for an input signal, $x_{dB}[n]$, less than threshold T based on ratio R using the following relationship:

$$g_{sc}[n] = T + (x_{dB}[n] - T) \cdot R$$

18.7.1 Gate

An expander with a large ratio ($\infty : 1$) is called a **gate**. This DR processor reduces the amplitude of an input signal to 0 when the amplitude is less than a threshold. A gate behaves as "open/closed" or "on/off" when the signal is above/below the threshold. A diagram describing the static characteristics of a gate is shown in Figure 18.21.

In Example 18.13, an implementation of an expander is demonstrated. The result of the processing is shown in Figure 18.22. Setting the ratio parameter to be relatively low values creates an expander (e.g., 2:1, 3:1). Setting the ratio parameter to higher values effectively creates a gate (e.g., 10:1, 20:1).

Examples: Create and run the following m-files in MATLAB.

```
% EXPANDEREXAMPLE
% This script demonstrates an expander/gate DR processor
%
```

```
% See also EXPANDER, COMPRESSOREXAMPLE
clear; clc; close all;


% Drums sound file
[in,Fs] = audioread('monoDrums.wav');


% Parameters for compressor
T = -20;    % Threshold = -20 dBFS
R = 3;      % Ratio = 3:1


% Initialize separate attack and release times
attackTime = 0.005;  % time in seconds
releaseTime = 0.4;   % time in seconds

out = expander(in,Fs,T,R,attackTime,releaseTime);
sound(out,Fs);
% sound(in,Fs); % For comparison
```

Example 18.13: Script: `expanderExample.m`

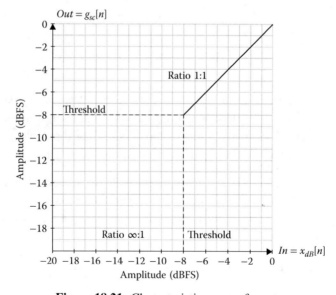

Figure 18.21: Characteristic curve of a gate

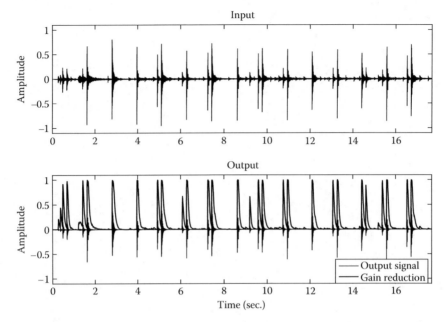

Figure 18.22: Waveform and gain reduction for signal processed by an expander

```
% EXPANDER
% This function implements an expander/gate DR processor.
% A similar approach is used as for a compressor, except
% the static characteristics are calculated differently.
%
% Input Variables
%    T : threshold relative to 0 dBFS
%    R : ratio (R to 1)
%    attackTime : units of seconds
%    releaseTime : units of seconds
%
% See also EXPANDER, COMPRESSOREXAMPLE
function [y] = expander(x,Fs,T,R,attackTime,releaseTime)
N = length(x);
y = zeros(N,1);
lin_A = zeros(N,1);

% Calculate separate attack and release times
```

```
alphaA = exp(-log(9)/(Fs * attackTime));
alphaR = exp(-log(9)/(Fs * releaseTime));


gainSmoothPrev = -144; % Initialize smoothing variable

% Loop over each sample to see if it is below thresh
for n = 1:N
    % Turn the input signal into a unipolar signal on the dB scale
    x_uni = abs(x(n,1));
    x_dB = 20*log10(x_uni/1);
    % Ensure there are no values of negative infinity
    if x_dB < -144
        x_dB = -144;
    end


    % Static characteristics
    if x_dB > T
        gainSC = x_dB; % Do not perform expansion
    else
        % Expander Calculation
        gainSC = T + (x_dB - T)*R ; % Perform downwards expansion

        % Gating (use instead of expander)
        %gainSC = -144;
    end


    gainChange_dB = gainSC - x_dB;


    % Smooth over the gainChange
    if gainChange_dB > gainSmoothPrev
        % attack mode
        gainSmooth = ((1-alphaA)*gainChange_dB) ...
            +(alphaA*gainSmoothPrev);
    else
        % release
        gainSmooth = ((1-alphaR)*gainChange_dB) ...
            +(alphaR*gainSmoothPrev);
    end


    % Convert to linear amplitude scalar
```

```
        lin_A(n,1) = 10^(gainSmooth/20);

        % Apply linear amplitude to input sample
        y(n,1) = lin_A(n,1) * x(n,1);

        % Update gainSmoothPrev used in the next sample of the loop
        gainSmoothPrev = gainSmooth;
end
t = [0:N-1]/Fs; t = t(:);

subplot(2,1,1);
plot(t,x); title('Input Signal');axis([0 14 -1.1 1.1]);
subplot(2,1,2);
plot(t,y,t,lin_A); title('Output'); axis([0 14 -1.1 1.1]);
legend('Output Signal','Gain Reduction');
```

Example 18.14: Function: `expander.m`

18.8 Upwards Compressor and Expander

A different approach to processing the dynamic range of a signal is to *increase* the amplitude of the input signal rather than decrease the amplitude in the conventional effects.

An **upwards compressor** increases the amplitude of an input signal when it is below the threshold. This has the effect of decreasing, or compressing, the dynamic range because there is a reduction in the difference between the maximum and minimum amplitude.

A diagram describing the static characteristics of an upwards compressor is shown in Figure 18.23. In this example, the ratio of 0.5:1 represents that for every 1 dB the input signal is below the threshold, the output is only 0.5 dB less than the threshold.

Conversely, an **upwards expander** increases the amplitude of an input signal when it is above the threshold. In this case, the dynamic range is increased, or expanded, because there is a greater difference between the maximum and minimum amplitude.

A diagram describing the static characteristics of an upwards expander is shown in Figure 18.24. In this example, the ratio of 0.5:1 represents that for every 0.5 dB the input signal is above the threshold, the output is 1 dB greater than the threshold.

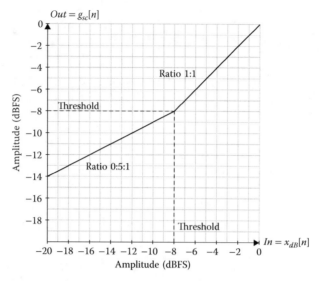

Figure 18.23: Characteristic curve of an upwards compressor

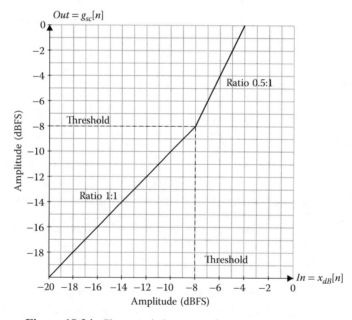

Figure 18.24: Characteristic curve of an upwards expander

18.9 Appendix I: De-Esser Effect

A **de-esser effect** is an extension of the conventional DR compressor effect by incorporating spectral filtering in the detection path (Figure 18.25). This makes it possible for the gain

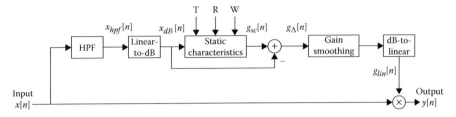

Figure 18.25: Block diagram for a de-esser effect

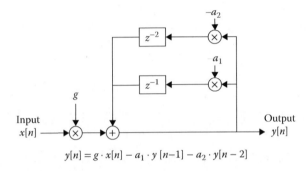

$$y[n] = g \cdot x[n] - a_1 \cdot y[n-1] - a_2 \cdot y[n-2]$$

Figure 18.26: Block diagram of an example second-order system

reduction to occur only when a desired narrow frequency range triggers the compressor, instead of the wide-band frequency range.

For the de-esser effect, an HPF is used in the detection path so the compressor applies gain reduction for high-frequency, sibilant vocal sounds like *-ess*, *-esh*, etc. Assuming the HPF cutoff frequency is set correctly, the compressor does not apply gain reduction for other low-frequency vocal sounds. Therefore, the de-esser can remove harsh vocal sounds without changing the rest of the signal.

18.10 Appendix II: Second-Order Control Systems

The gain smoothing filter used in DR processors can be implemented as a second-order system. In Figure 18.8, a first-order system is presented and can be extended to a second-order system by adding an additional feedback path with two samples of delay, as shown in Figure 18.26. Additionally, the gain smoothing filter could be implemented in the general form of a bi-quad filter (Section 13.3.2).

The step response of a first-order system resembles a simple exponential curve (Figure 18.9). By increasing the order of the system, the step response can resemble an exponential curve, as well as a damped sinusoid (Figure 18.27).

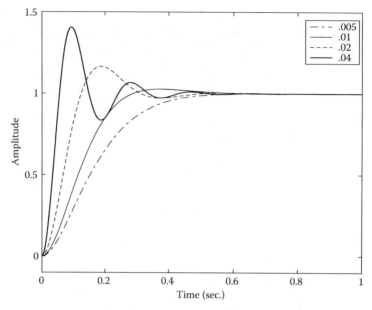

Figure 18.27: Step response of second-order systems

18.10.1 Characteristics of a Second-Order Control System

The step response of a second-order system is defined by several characteristics. Similar to a first-order system, **rise time** is defined as the time for the step response to transition from 10% to 90% of the step input, $t_r = t_{90} - t_{10}$. The **steady-state value** of the step response is the value of the output as time approaches infinity, $y_{ss} = \lim_{n \to \infty} y[n]$. The **percent overshoot** describes the amount by which the peak response exceeds the steady-state value, $\%OS = \frac{y_{pk} - y_{ss}}{y_{ss}} \cdot 100$. The **settling time**, t_s, is the time for the step response to reach and stay within 2% of the steady-state value, y_{ss}. These characteristics are described visually in Figure 18.28.

18.10.2 Bi-Quad Parameters and Step Response Characteristics

If a bi-quad LPF is used as the second-order system, its conventional parameters (Q and cutoff frequency) can be used to change the characteristics of the step response. In Figure 18.29, several examples are plotted for various Q values while holding the cutoff frequency constant. By increasing the resonance of the LPF, the percent overshoot can be changed.

Conversely, several examples are plotted in Figure 18.30 for various cutoff frequencies while holding the Q value constant. By increasing the filter's cutoff frequency, the rise time and settling time are decreased.

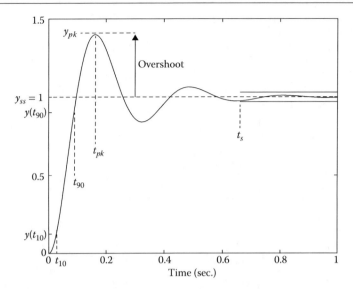

Figure 18.28: Step response characteristics of a second-order system

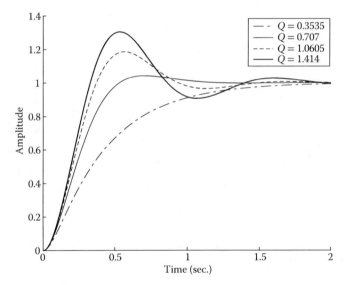

Figure 18.29: Step response of a bi-quad LPF with several Q values

Examples: Create and run the following m-file in MATLAB.

```
% BIQUADSTEP
% This script demonstrates the result of taking the
% step response of a bi-quad LPF. Examples include
% changing the cutoff frequency and Q.
%
```

```
% See also BIQUADFILTER

clc;clear;close all;
% Input signal
Fs = 48000; Ts = 1/Fs;
x = ones(2*Fs,1); % Step input
N = length(x);
t = [0:N-1] * Ts; t = t(:);

% Changing the cutoff frequency
Q = 1.414;
dBGain = 0;
figure(1); hold on;
for freq = 1:4
    y = biquadFilter(x,Fs,freq,Q,dBGain,'lpf',1);
    plot(t,y);
end
hold off;legend('f = 1','f = 2','f = 3','f = 4');
xlabel('Time (sec.)');

% Changing the bandwidth Q
freq = 1;
dBGain = 0;
figure(2); hold on;
for Q = 0.707/2:0.707/2:1.414
    y = biquadFilter(x,Fs,freq,Q,dBGain,'lpf',1);
    plot(t,y);
end
hold off;legend('Q = 0.3535','Q = 0.707','Q = 1.0605','Q = 1.414');
xlabel('Time (sec.)');
```

Example 18.15: Script: `biquadStep.m`

18.10.3 *Second-Order System Design Function*

A MATLAB function is included with the book's *Additional Content* to design a second-order system based on the step response characteristics: percent overshoot, peak time, and settling time. This function returns the coefficients of a bi-quad filter that fulfills the design specifications. The syntax to use the function is $[b,a]$ = stepDesign(Fs,OS,T,

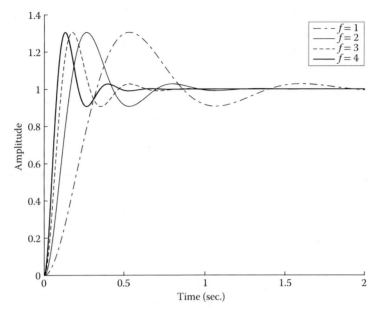

Figure 18.30: Step response of a bi-quad LPF with several cutoff frequencies

'type'), where Fs is the sampling rate, OS is the percent overshoot, and T is the time in seconds for the characteristic 'type', either 'pk' for peak time or 'st' for settling time.

Examples: Create and run the following m-file in MATLAB.

```
% STEPDESIGNEXAMPLE
% This script demonstrates the stepDesign function
% for designing a second-order system with
% specified characteristics. The system is designed
% based on the percent overshoot, setttling time,
% and peak time.
%
% See also STEPDESIGN
clc;clear;close all;

Fs = 48000;

% Example - filter design based on settling time
OS = 20;    % Percent overshoot
% Settling time in seconds
ts = .25;    % (within 2% of steady-state)
```

```
[b,a] = stepDesign(Fs,OS,ts,'st');
n = 1*Fs;  % number of seconds for step response
[h,t] = stepz(b,a,n,Fs);
plot(t,h);

figure;

% Example - filter design based on peak time
OS = 10;   % Percent overshoot
tp = 0.75; % Peak time in seconds
[b,a] = stepDesign(Fs,OS,tp,'pk');
n = 2*Fs;  % number of seconds for step response
[h,t] = stepz(b,a,n,Fs);
plot(t,h);
```

Example 18.16: Script: `stepDesignExample.m`

Bibliography

Giannoulis, D., Massberg, M., & Reiss, J. (2012). Digital dynamic range compressor design—a tutorial and analysis. *AES: Journal of the Audio Engineering Society. 60*(6), 399–408.

Index